Flow Perturbation Gas Chromatography

CHROMATOGRAPHIC SCIENCE

A Series of Monographs

Editor: JACK CAZES
Sanki Laboratories, Inc.
Sharon Hill, Pennsylvania

Volume 1: Dynamics of Chromatography
J. Calvin Giddings

Volume 2: Gas Chromatographic Analysis of Drugs and Pesticides
Benjamin J. Gudzinowicz

Volume 3: Principles of Adsorption Chromatography: The Separation of Nonionic Organic Compounds (out of print)
Lloyd R. Snyder

Volume 4: Multicomponent Chromatography: Theory of Interference (out of print)
Friedrich Helfferich and Gerhard Klein

Volume 5: Quantitative Analysis by Gas Chromatography
Joseph Novák

Volume 6: High-Speed Liquid Chromatography
Peter M. Rajcsanyi and Elisabeth Rajcsanyi

Volume 7: Fundamentals of Integrated GC-MS (in three parts)
Benjamin J. Gudzinowicz, Michael J. Gudzinowicz, and Horace F. Martin

Volume 8: Liquid Chromatography of Polymers and Related Materials
Jack Cazes

Volume 9: GLC and HPLC Determination of Therapeutic Agents (in three parts)
Part 1 edited by Kiyoshi Tsuji and Walter Morozowich
Part 2 and 3 edited by Kiyoshi Tsuji

Volume 10: Biological/Biomedical Applications of Liquid Chromatography
Edited by Gerald L. Hawk

Volume 11: Chromatography in Petroleum Analysis
Edited by Klaus H. Altgelt and T. H. Gouw

Volume 12: Biological/Biomedical Applications of Liquid Chromatography II
Edited by Gerald L. Hawk

Volume 13: Liquid Chromatography of Polymers and Related Materials II
Edited by Jack Cazes and Xavier Delamare

Volume 14: Introduction to Analytical Gas Chromatography: History, Principles, and Practice
John A. Perry

Volume 15: Applications of Glass Capillary Gas Chromatography
Edited by Walter G. Jennings

Volume 16: Steroid Analysis by HPLC: Recent Applications
Edited by Marie P. Kautsky

Volume 17: Thin-Layer Chromatography: Techniques and Applications
Bernard Fried and Joseph Sherma

Volume 18: Biological/Biomedical Applications of Liquid Chromatography III
Edited by Gerald L. Hawk

Volume 19: Liquid Chromatography of Polymers and Related Materials III
Edited by Jack Cazes

Volume 20: Biological/Biomedical Applications of Liquid Chromatography IV
Edited by Gerald L. Hawk

Volume 21: Chromatographic Separation and Extraction with Foamed Plastics and Rubbers
G. J. Moody and J. D. R. Thomas

Volume 22: Analytical Pyrolysis: A Comprehensive Guide
William J. Irwin

Volume 23: Liquid Chromatography Detectors
Edited by Thomas M. Vickrey

Volume 24: High-Performance Liquid Chromatography in Forensic Chemistry
Edited by Ira S. Lurie and John D. Wittwer, Jr.

Volume 25: Steric Exclusion Liquid Chromatography of Polymers
Edited by Josef Janča

Volume 26: HPLC Analysis of Biological Compounds: A Laboratory Guide
William S. Hancock and James T. Sparrow

Volume 27: Affinity Chromatography: Template Chromatography of Nucleic Acids and Proteins
Herbert Schott

Volume 28: HPLC in Nucleic Acid Research: Methods and Applications
Edited by Phyllis R. Brown

Volume 29: Pyrolysis and GC in Polymer Analysis
Edited by S. A. Liebman and E. J. Levy

Volume 30: Modern Chromatographic Analysis of the Vitamins
Edited by Andre P. De Leenheer, Willy E. Lambert, and Marcel G. M. De Ruyter

Volume 31: Ion-Pair Chromatography
Edited by Milton T. W. Hearn

Volume 32: Therapeutic Drug Monitoring and Toxicology by Liquid Chromatography
Edited by Steven H. Y. Wong

Volume 33: Affinity Chromotography: Practical and Theoretical Aspects
Peter Mohr and Klaus Pommerening

Volume 34: Reaction Detection in Liquid Chromatography
Edited by Ira S. Krull

Volume 35: Thin-Layer Chromatography: Techniques and Applications, Second Edition, Revised and Expanded
Bernard Fried and Joseph Sherma

Volume 36: Quantitative Thin-Layer Chromatography and Its Industrial Applications
Edited by Laszlo R. Treiber

Volume 37: Ion Chromatography
Edited by James G. Tarter

Volume 38: Chromatographic Theory and Basic Principles
Edited by Jan Åke Jönsson

Volume 39: Field-Flow Fractionation: Analysis of Macromolecules and Particles
Josef Janča

Volume 40: Chromatographic Chiral Separations
Edited by Morris Zief and Laura J. Crane

Volume 41: Quantitative Analysis by Gas Chromatography, Second Edition, Revised and Expanded
Josef Novak

Volume 42: Flow Perturbation Gas Chromatography
N. A. Katsanos

Volume 43: Ion Exchange Chromatography of Proteins
Shuichi Yamamoto, Kazuhiro Nakanishi, and Ryuichi Matsuno

Volume 44: Countercurrent Chromatography: Theory and Practice
Edited by N. Bhushan Mandava and Yoichiro Ito

Flow Perturbation Gas Chromatography

N. A. Katsanos
University of Patras
Patras, Greece

Marcel Dekker, Inc.   New York • Basel

Library of Congress Cataloging-in-Publication Data

N. A. Katsanos
Flow perturbation gas chromatography.

(Chromatographic science ; v. 42)
Includes bibliographies and index.
1. Gas Chromatography. I. Title. II. Series.
QD79.C45K38 1988 543'.0897 87-30487
ISBN 0-8247-7833-2

MARCEL DEKKER, INC.
270 Madison Avenue, New York, New York 10016

Current printing (last digit):
10 9 8 7 6 5 4 3 2 1

PRINTED IN THE UNITED STATES OF AMERICA

Preface

Gas chromatography (GC) is used today mainly for chemical analysis, but there is an increasing interest in recent years to employ it as a means for physicochemical measurements. Indeed GC offers many possibilities for such measurements, leading in many cases to very precise and accurate results, with relatively cheap instrumentation and a very simple experimental setup. Two books have been published (1,2) dealing exclusively with physicochemical measurements. However, the methods described in these books are based on the traditional techniques of GC, namely, elution, frontal analysis, and displacement development, under constant gas flowrate. Since 1967, two methods for physicochemical measurements have been developed that are based on perturbations imposed on the carrier gas flow: the stopped-flow technique, introduced in 1967 by C. S. G. Phillips and co-workers (3), and the reversed-flow technique, invented in 1980 by us (4) and reviewed in 1984 (5). This book discusses these two new methods in detail, both from the theoretical and the practical point of view. It is my belief that chromatography requires both approaches if it is to be of any value to research workers. No artificial barrier should be inserted between theory and practice.

The book is directed primarily to two groups of workers: first, to those who need a simple and fairly accurate method for the determination of a certain physicochemical quantity, like diffusion coefficients in gases, molecular diameters and critical volumes in gases, mass transfer coefficients, activity coefficients in liquid mixtures, adsorption-desorption rates, rate of action of gases on solids, adsorption equilibrium constants, rate constants and activation parameters, as well as conversion yields, of a variety of surface-catalyzed chemical reactions. These last applications interest all people working in the field of heterogeneous catalysis.

The second group includes researchers working on the development of gas chromatography from the point of view of theoretical advances, as well as the introduction of new methods and techniques.

In Chapter 1 the gas chromatographic method is briefly exposed to create the necessary background for the subsequent discussion. Here the various chromatographic terms and notations, together with the basic equations of gas chromatography, are given. A short account of physicochemical measurements by the existing conventional techniques follows, and the chapter ends with an introduction to the carrier gas flow perturbations.

Chapter 2 deals with the first flow perturbation method, namely the stopped-flow technique. After a brief introduction, various applications are directly described both theoretically and practically.

The next four chapters are devoted to the newest flow perturbation method, viz., the reversed-flow technique. Chapter 3 comprises a general exposition of the experimental setup and the general theoretical analysis of the method. For those interested in immediate and direct applications, only the final few equations are needed. Chapters 4, 5, and 6 describe the measurement of various physicochemical quantities, classified according to three limiting cases of experimental arrangement. These are: (a) an empty diffusion column, (b) a filled diffusion column, and (c) a filled sampling column.

Throughout the book schematic figures are used as a guide to the various arrangements.

I am indebted to Mrs. Margaret Barkoula and Mrs. Anna Sinou-Karahaliou for their kind help in typing the book. Most of the drawings have been made by Mrs. Georgia Paraschi and her talent is apparent in them. The credit for some types of technical assistance goes to Mr. G. Pittas.

N. A. Katsanos

REFERENCES

1. R. J. Laub and R. L. Pecsok, Physicochemical Applications of Gas Chromatography, Wiley, New York, 1978.
2. J. R. Conder and C. L. Young, Physicochemical Measurement by Gas Chromatography, Wiley, Chichester, 1979.
3. C. S. G. Phillips, A. J. Hart-Davis, R. G. L. Saul, and J. Wormald, J. Gas Chromatogr., 5:424 (1967).
4. N. A. Katsanos and I. Georgiadou, J. Chem. Soc., Chem. Commun. 242 (1980).
5. N. A. Katsanos and G. Karaiskakis, Adv. Chromatogr., 24: 125 (1984).

Contents

Preface iii

Chapter 1 The Gas Chromatographic Method 1

I. Introduction 1
II. Terminology 2
III. Chromatographic Methods 4
IV. The Basic Equations of Gas Chromatography 7
V. Physicochemical Measurements by Gas Chromatography 30
VI. The Carrier Gas Flowrate Perturbation 34
List of Symbols 36
References 38

Chapter 2 The Stopped-Flow Technique 41

I. Introduction 41
II. Experimental Setup 47
III. Ideal Stopped-Flow Gas Chromatography with Chemical Reactions in the Stationary Phase 49
IV. Nonideal Stopped-Flow Gas Chromatography 72
V. Measurement of Diffusion Coefficients in Gases 78
List of Symbols 82
References 84

Chapter 3 The Reversed-Flow Technique 87

I. Introduction 87
II. General Experimental Setup 88
III. Reversed-Flow GC as a Sampling Technique 92
IV. The Chromatographic Sampling Equation 95
V. Calculation of Rate Coefficients and Equilibrium Constants from the Sample Peaks 107
List of Symbols 108
References 110

Chapter 4 Reversed-Flow with an Empty Diffusion Column 113

I. Gas Diffusion Coefficients in Binary Mixtures 113
II. Diffusion Coefficients in Multicomponent Gas Mixtures 138
III. Relative Molar Responses, Collision Diameters, and Critical Volumes of Gases 142
IV. Lennard-Jones Parameters 154
List of Symbols 158
References 160

Chapter 5 Reversed-Flow with a Filled Diffusion Column 163

I. Introduction 164
II. Mass Transfer Across a Gas-Liquid Boundary 165
III. Mass Transfer Coefficients for the Evaporation of Liquids and Diffusion Coefficients of Vapors 180
IV. Mass Transfer Across Phase Boundaries Near the Junction of the Diffusion and the Sampling Column 203
V. Mass Transfer Across Phase Boundaries Extended to All the Diffusion Column 212
List of Symbols 221
References 223

Chapter 6 Reversed-Flow with a Filled Sampling Column 225

I. Introduction 226
II. Kinetics of Surface-Catalyzed Reactions 226
III. Reaction Kinetics with the Pulse Technique 236
IV. Reaction Kinetics with the Diffusion Technique 260

V. Mass Transfer Coefficients of Vapors on Solids 285
List of Symbols 293
References 294

Index 297

Flow Perturbation Gas Chromatography

1

The Gas Chromatographic Method

I. Introduction

II. Terminology

III. Chromatographic Methods

IV. The Basic Equations of Gas Chromatography

- A. Retention parameters
- B. Experimental determination of retention volumes
- C. Broadening factors in gas chromatography
- D. Theoretical plates and van Deemter equation

V. Physicochemical Measurements by Gas Chromatography

VI. The Carrier Gas Flowrate Perturbation

List of Symbols

References

I. INTRODUCTION

Chromatography as a phenomenon was discovered at the beginning of this century by the Russian botanist Tswett (1), and was first used by him for isolation of plant pigments. The work was not appreciated immediately and was forgotten for about 25 years. However, after "reinvention" of the method by Kuhn and

Lederer in 1931(2,3), liquid chromatography was fully developed in less than 15 years, including thin-layer and paper chromatography. The chromatographic process was first described mathematically by Wilson in 1940, assuming solute adsorption-desorption equilibria (4). The first mention of physicochemical measurements by gas chromatography (GC) was in 1947, when Glueckauf (5) pointed out the possibility of adsorption isotherms determination by gas-solid chromatography (GSC). Five years later, James and Martin published a classic paper (6) introducing gas-liquid chromatography (GLC). In less than 10 years, this mehod became one of the most widely used analytical techniques. A compilation of personal recollections of the pioneers involved in the development of chromatography has been edited by Ettre and Zlatkis (7).

There are so many excellent textbooks on gas chromatography today, to mention only a few (8-11), that any extensive exposure of the gas chromatographic method seems unnecessary. We therefore limit this chapter only to those matters immediately connected to physicochemical measurements and to the following chapters.

II. TERMINOLOGY

All chromatographic methods are separation methods belonging to two main categories; adsorption chromatography and partition chromatography. In the same way that fractional distillation is based on repeated evaporations and condensations of a liquid mixture, adsorption chromatography depends on successive adsorptions and desorptions of the components of a liquid or gas mixture on a solid adsorbent. Similarly, partition chromatography is based on repeated distributions of the mixture components between two liquid phases in contact or between a gaseous and a liquid phase.

The analogy between distillation and chromatography can be pushed a litle further by reminding that in the first method, two phases move relative to each other, a stream of ascending vapor and a stream of descending liquid, to effect the repeated evaporation and condensation. Similarly, in chromatography for a repetition of the distribution of a solute component between two phases, it is necessary that the two phases move relative to each other. The simplest possible way to do this is to have the one phase fixed, usually in a narrow tube (chromatographic column) and the other phase streaming through the tube as a liquid or a gas. The first is called the stationary phase and the second the mobile phase. In adsorption chromatography, the stationary phase is a solid adsorbent, whereas in partition chromatography, it is a liquid held on a porous solid support or on the walls of a capillary chromatographic column. In liquid chromatography the mobile phase is a liquid, whereas in gas chromatography, it is a gas (carrier gas). Combining each of the two kinds of the stationary phase with each of the two kinds of the mobile phase, one obtains the four basic types of chromatography. These are: 1. liquid-solid chromatography (LSC), 2. liquid-liquid chromatography (LLC), 3. gas-solid chromatography (GSC), and 4. gas-liquid chromatography (GLC), the first noun denoting the mobile and the second the stationary phase.

Apart from these basic types, there are some other categories of liquid chromatography with stationary phases like an ion-exchange resin (ion-exchange chromatography), a gel (gel-permeation chromatography), biological macromolecules (affinity chromatography), complex cellulose/water materials (paper chromatography), an adsorbent, liquid, gel, or ion-exchange resin, not held in a tubular column, but spread as a thin layer on a piece of glass (thin-layer chromatography). There is also the special type of supercritical fluid chromatography, in which the mobile phase is a vapor just above critical point.

Two terms widely used in the literature are ideal and nonideal chromatography. In the first (also called equilibrium chromatography), the equilibration of the solute components of the mixture under separation between the mobile and the stationary phase is assumed instantaneous. In nonideal (nonequilibrium) chromatography there is a finite rate for equilibration of the solutes between the two phases owing to various "resistances," like slow adsorption-desorption phenomena, intraparticle diffusion, small mass transfer coefficients across the phase boundaries, and so forth.

Another pair of terms is linear and nonlinear chromatography, meaning that the law governing equilibration of the solutes between the mobile and the stationary phase is a linear one, like Henry's law and a linear adsorption isotherm, or a nonlinear one, like Langmuir's adsorption isotherm.

III. CHROMATOGRAPHIC METHODS

The mixture to be analyzed and separated into its components may be a gas, a liquid, or even a solid, provided of course that at the separation temperature the components are stable and have a sufficient vapor pressure (about 0.1 torr).

Three main techniques are used in chromatography: 1. elution development, 2. frontal analysis, and 3. displacement development. In the first technique, the mixture under separation is introduced into the chromatographic column usually as a "pulse" of gas or vapor, i.e., as a narrow distribution of concentration versus time or versus length coordinate along the column. Then, by passing pure mobile phase (e.g., carrier gas) through the column, the components emerge from the other end of the column as separate peaks, like those of Fig. 1.1. In the case of linear-ideal chromatography these peaks have a gaussian shape. Mathematically, the ideal chromatographic

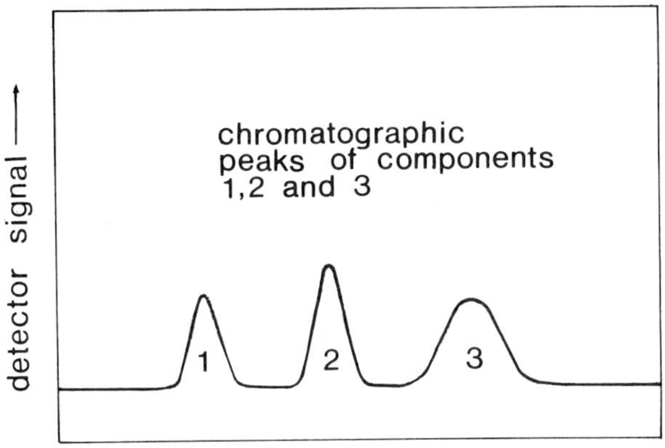

Fig. 1.1 The appearance of an elution development chromatogram.

column is a so-called gaussian operator; i.e., each element of the input distribution profile of known area is broadened to a gaussian distribution of the same area by passage through the column. In principle, it is possible to calculate the shape of the output peak for any shape of input function. For a thorough presentation of this point see Ref. (12).

In frontal analysis, the mixture (liquid or gas) is continuously fed into the clumn under constant concentration in the mobile phase. Once the stationary phase has been saturated in the mixture components, this then flows through the column with its original composition. The use of the technique consists of measuring the concentration profile of the front leaving the column, or the break-through curve, as it is called. The least retained component breaks through first, and it is the only one to be obtained in a pure form. The appearance of a frontal analysis chromarogram is depicted in Fig. 1.2.

Finally, in displacement development a discrete sample of the mixture to be analyzed is placed at the beginning of the chromatographic column and a carrier gas stream loaded with

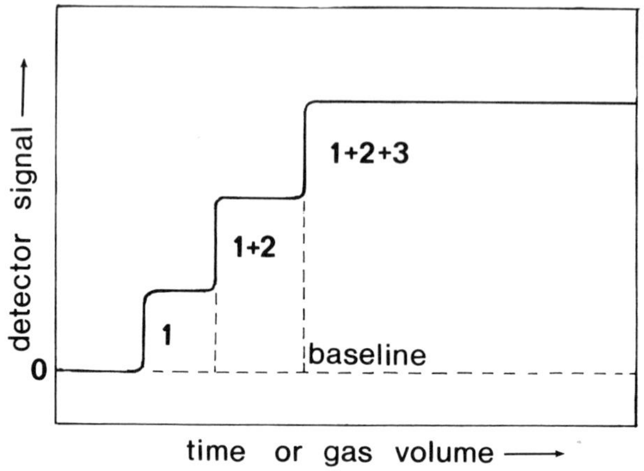

Fig. 1.2 The shape of a frontal analysis chromatogram with components 1, 2, and 3.

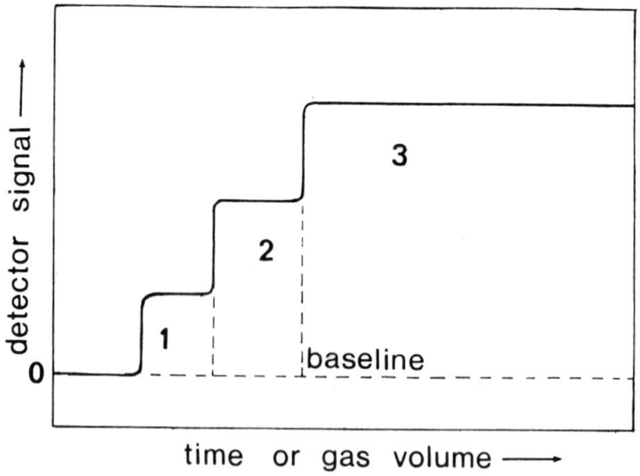

Fig. 1.3 Displacement development chromatogram.

a constant concentration of a "displacer," i.e., a substance more strongly adsorbed than the components of the mixture, is passed through the column. A series of steps is produced each corresponding to a pure component of the mixture, as shown in Fig. 1.3. The component bands, however, are not separated by a region of pure mobile phase, as in elution development.

IV. THE BASIC EQUATIONS OF GAS CHROMATOGRAPHY

A. Retention Parameters

The quantitative theory of gas chromatography is based on a simple first order partial differential equation. As was mentioned earlier, a form (oversimplified) of the equation was first given by Wilson (4). This was modified by DeVault (13) to give the so-called first-order conservation equation:

$$\frac{\partial c_G}{\partial x} + a_G \frac{\partial c_G}{\partial V} + m_S \frac{\partial c_S}{\partial V} = 0 \qquad (1\text{-}1)$$

where

c_G = concentration of a solute vapor in the gas phase of the column (mol cm^{-3})

c_S = concentration of the solute in the stationary phase (mol g^{-1})

a_G = volume of gas phase per unit length of column, or cross-sectional area of void space (cm^2)

m_S = mass of stationary phase per unit length of column (g cm^{-1})

x = distance coordinate along the column (cm)

V = volume of carrier gas passed through the column (cm^3)

Various other forms of this equation have been published. A simple derivation is given here based on the time, t, as the independent variable rather than the volume, V, of the carrier gas. The derivation can be followed by referring to Fig. (1.4).

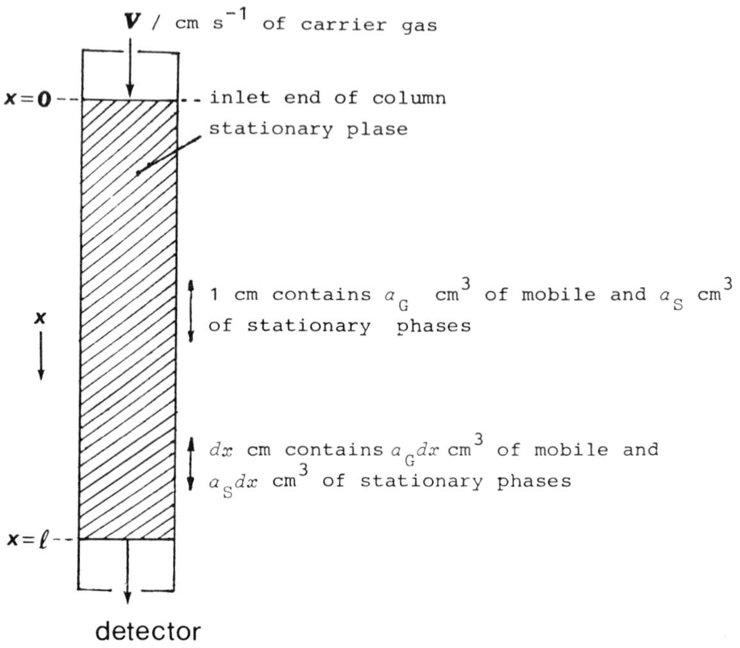

Fig. 1.4 Schematic representation of the chromatographic column.

The sample under separation is introduced at the inlet end of column (x = 0) in vapor form. To simplify things, we assume that this sample contains only one component, which is distributed instantaneously inside the column between the mobile and the stationary phases, so that its concentrations are c_G and c_S (mol cm^{-3}), respectively. These concentrations change with time and distance along the column; i.e., $c_G = c_G(x,t)$ and $c_S = c_S(x,t)$. We write the mass balance of this single component for an infinitesimal length, dx, in which there are $a_G dx$ cm^3 of mobile and $a_S dx$ cm^3 of stationary phase. The rate of change of the component concentrations with time, $(\partial c_G/\partial t)/$ mol $\cdot$ cm^{-3} $\cdot$ s^{-1} and $(\partial c_S/\partial t)/$mol $\cdot$ cm^{-3}s^{-1}, in the length, dx multiplied by the volumes of the two phases, $a_G dx/\text{cm}^3$ and $a_S dx/\text{cm}^3$, gives the rate of change of the total amount, $a_G dx$

$(\partial c_G/\partial t)$/mol · s^{-1} and $a_S dx(\partial c_S/\partial t)$/mol· s^{-1}, of the component in the mobile and stationary phase, respectively. The sum of these two rates must be equal to the amount of the component entering the section, dx, per unit time, $a_G v c_G$/mol · s^{-1}, v/cm· s^{-1} being the linear velocity of the mobile phase (carrier) gas) minus the amount per unit time $a_G v(c_G + dc_G)$/mol · s^{-1} leaving the section, with dc_G denoting the change of c_G due to the passage through dx. Thus

$$a_G dx \frac{\partial c_G}{\partial t} + a_S dx \frac{\partial c_S}{\partial t} = a_G v c_G - a_G v(c_G + dc_G)$$

Canceling $a_G v c_G$ on the right-hand side of this equation, and dividing throughout by $a_G dx$, we obtain

$$\frac{\partial c_G}{\partial t} + r\frac{\partial c_S}{\partial t} = -v\frac{\partial c_G}{\partial x} \qquad (1\text{-}2)$$

where $r = a_S/a_G$ is the volume ratio of stationary and mobile phases, and dc_G/dx has been replaced by the partial derivative $\partial c_G/\partial x$ to take into acount that c_G is a function of two independent variables, x and t. The derivation of Eq. (1-2) was done under two assumptions: 1. axial diffusion of the component in the gas phase is negligible, which is not unrealistic for high enough linear velocities, v, and 2. equilibration of the component between the two phases is instantaneous; i.e., chromatography is ideal.

Only a little step is required to make Eq. (1-2) coincide with the DeVault equation (1-1). This is to replace ∂t by $\partial V/a_G v$, since

$$V = a_G vt \quad \text{and} \quad dV = a_G v dt$$

and finally divide throughout by v. Thus, the only differences between the two equations are that (1-2) has the time, t, instead of the volume, V, as independent variable, and also a_S in place of m_S, i.e., the volume of stationary phase instead of its mass per unit length of column.

Equation (1-2) is a partial differential equation and its solution depends on: 1. the relation between c_G and c_S, and 2. the initial and the boundary conditions. In linear chromatography, the relation between c_G and c_S is a linear isotherm:

$$c_S = Kc_G \tag{1-3}$$

where K is the partition coefficient or distribution constant (here dimensionless). Substitution of Kc_G for c_S in Eq. (1-2) gives, after some rearrangement:

$$(1 + k)\frac{\partial c_G}{\partial t} = -v\frac{\partial c_G}{\partial x} \tag{1-4}$$

where k is given by the relation

$$k = Kr \tag{1-5}$$

and is termed partition ratio. It expresses the ratio of the amount of the solute component per unit length of the stationary phase to the solute amount per unit length of the mobile phase.

The initial condition is

$$c_G(x,0) = c_S(x,0) = 0 \tag{1-6}$$

since at t = 0 the column is empty of any solute component along its entire length.

As regards the boundary condition at x = 0, this depends on the mode of solute introduction onto the chromatographic column. We shall solve Eq. (1-4) for the most common input distribution; i.e., an instantaneous injection of a pulse of solute. Such a distribution is described mathematically by Dirac's delta function $\delta(t)$, which is generally written $\delta(t - a)$ with $a \geq 0$ and has a non-zero value only for $t = a$ (here $t = 0$), whereas $\delta(t - a) = 0$ for $t \neq a$ (here for $t \neq 0$). Thus the boundary condition at $x = 0$ is

$$c_G(o,t) = \frac{m}{a_G v} \delta(t) \tag{1-7}$$

where m is the amount (mol) of the component to be analyzed.

Equation (1-4) can now be integrated by using Laplace transformation [14] of its terms with respect to t, under the initial condition (1-6):

$$(1 + k)[pC_G - c_G(x,0)] = -v \frac{dC_G}{dx}$$

or

$$\frac{dC_G}{dx} = -\frac{(1 + k)p}{v} C_G \tag{1-8}$$

where p is the transform parameter, and the capital letter C_G denotes the Laplace transform of c_G with respect to t. It is noteworthy that the above transformation has changed a partial differential equation like (1-4) into an ordinary differential equation like (1-8), having only one independent variable, i.e., x. This equation can easily be solved by separation of variables and integration, the result being

$$C_G = I \exp\left[-\frac{(1 + k)x}{v} p\right] \tag{1-9}$$

where I is the constant of integration. It can be evaluated with the help of the boundary condition (1-7), since for $x = 0$ Eq. (1-9) gives $C_G = I$, and the Laplace transformation of (1-7) results in

$$c_G(0,p) = \frac{m}{a_G v}$$

Thus, $I = m/a_G v$ and substituting this for I in (1-9), followed by inverse Laplace transformation with respect to p, we obtain

$$c_G = \frac{m}{a_G v} \delta\left(t - \frac{1 + k}{v} x\right) \tag{1-10}$$

According to the properties of the delta function, the right-hand side of this equation has a non-zero value only for

$$t = \frac{1 + k}{v} x \tag{1-11}$$

This relation forms the basis for chromatographic analysis because it predicts that each solute component moves along the column with a certain linear velocity, $u = x/t$, which depends on the value of the partition ratio k:

$$u = \frac{x}{t} = \frac{v}{1 + k} \tag{1-12}$$

Thus, if two or more components 1, 2, ... have different k values $k_1, k_2, \ldots$, their linear velocities $u_1, u_2, \ldots$ will be different, with the result: 1. they reach different points of the column $x_1, x_2, \ldots$ at the same time, since $x = ut$; 2. they reach the detector at the end of the column ($x = l$) at different times $t_1, t_2, \ldots$, since $t = l/u$. In case (1) one usually measures the ratio u/v, called retardation factor, R_f, representing the ratio of the distance traveled by the component to that traveled by the mobile phase, e.g., the solvent:

$$R_f = \frac{u}{v} = \frac{1}{1 + k} \leq 1 \tag{1-13}$$

This concept is frequently used in paper chromatography.

In case (2) above, we term the times $t_1, t_2, \ldots$, required by the component to travel the entire column length, l, retention times, $t_{R1}, t_{R2}, \ldots$. The relation between t_R and k is obtained from Eq. (1-11) by putting x = l:

$$t_R = \frac{l}{v}(1 + k) = t_M(1 + k) \qquad (1\text{-}14)$$

where $t_M = l/v$ is the so-called hold-up time, or dead time. It is the time required by the solvent or the carrier gas or any other unretained substance (k=0) to travel the whole column length l. The difference $t_R - t_M = kt_M$ is the adjusted retention time t'_R.

If Eq. (1-14) is multiplied by the volume flowrate, $\dot{V} = a_g v$, of the mobile phase, e.g., the carrier gas, through the column, one obtains the corrected retention volume V_R^o for each solute component:

$$V_R^o = \dot{V}t_R = \dot{V}t_M(1 + k) = V_M^o(1 + k) \qquad (1\text{-}15)$$

where V_M^o is the corrected hold-up volume, or dead volume. It is a measure of the volume of the solvent or carrier gas required to elute an unretained component. The net retention volume, V_N, is the difference $V_R^o - V_M^o$, which from Eq. (1-15) is seen to be equal to $V_M^o k$. Taking also into account Eq. (1-5), we obtain

$$V_N = V_M^o k = \dot{V}t_M K \frac{a_S}{a_G} = a_G v \frac{l}{v} K \frac{a_S}{a_G} = la_S K \qquad (1\text{-}16)$$

But la_S is the total volume, V_S, of the stationary phase (of the whole phase if this is a solid adsorbent, or only of the liquid if this is the stationary phase held on a solid support). Thus we arrive at the important relation:

$$V_N = V_R^o - V_M^o = V_S K \qquad (1\text{-}17)$$

If the concentration of the solute in the stationary phase, c_S, is expressed per unit mass of the stationary phase, the partition coefficient in the isotherm (1-3) would have the usual units cm^3g^{-1}, and Eq. (1-17) would become

$$V_N = W\beta \tag{1-18}$$

where W is the total weight in g of the stationary phase, and β the partition coefficient in cm^3g^{-1}.

B. Experimental Determination of Retention Volumes

The direct experimental data in gas chromatography are the volume flowrate, V_f, measured at the temperature of the flowmeter, T_f, and the chromatogram. The flowrate requires some corrections before it is used in the equations of the previous paragraph.

1. Temperature and Vapor Pressure Corrections

The first of these corrections is to bring the volume flowrate at the column temperature, T_c, multiplying by the factor T_c/T_f. In addition, the flowrate is often determined with a soap-film flowmeter, and this necessitates a correction for the vapor pressure of the soap slution, which is approximately equal to the vapor pressure, p_w, of pure water at the flowmeter temperature. If p_0 is the pressure at the flowmeter, the correction factor is $1 - p_w/p_0$. The above two corrections then are incorporated into the relation:

$$\dot{V}_c = \dot{V}_f \frac{T_c}{T_f}\left(1 - \frac{p_w}{p_0}\right) \tag{1-19}$$

2. Compressibility Correction

In gas chromatography where the mobile phase is a gas, there is a pressure gradient along the column with the result that the gas flowrate is not the same at all points along the column. The correction factor for the pressure drop is easily calculated [8], and is given by the relation:

$$j = \frac{3}{2} \cdot \frac{(p_i/p_0)^2 - 1}{(p_i/p_0)^3 - 1} \tag{1-20}$$

where p_i and p_0 are the pressures at the inlet end ($x = 0$) and the outlet end ($x = 1$) of the column, respectively. Multiplying $\dot{V}_c$ of Eq. (1-19) by j, we find the corrected volume flowrate, $\dot{V}$, of the carrier gas:

$$\dot{V} = j\dot{V}_c \tag{1-21}$$

This then is used in Eq. (1-15) to find the corrected retention volume, V_R^o, by multiplication with the retention time, t_R.

3. Calculation of t_R and t_M From the Chromatogram

A typical elution chromatogram is shown in Fig. 1.5. The retention time, t_R, of an injected solute and the dead time, t_M, are measured as indicated in the Fig. 1.5. Instead of measuring times, one can find t_R and t_M from the respective distances z_R and z_M on the recorder chart and divide them by the chart speed, v_c:

$$t_R = \frac{z_R}{v_c} \qquad t_M = \frac{z_M}{v_c} \tag{1-22}$$

Multiplying $\dot{V}$ from Eq. (1-21) with t_R or t_M, we find the corrected retention volume V_R^o, or the corrected hold-up volume V_M^o, respectively (Eq. [1-15]). Finally, the net retention volume V_N is found:

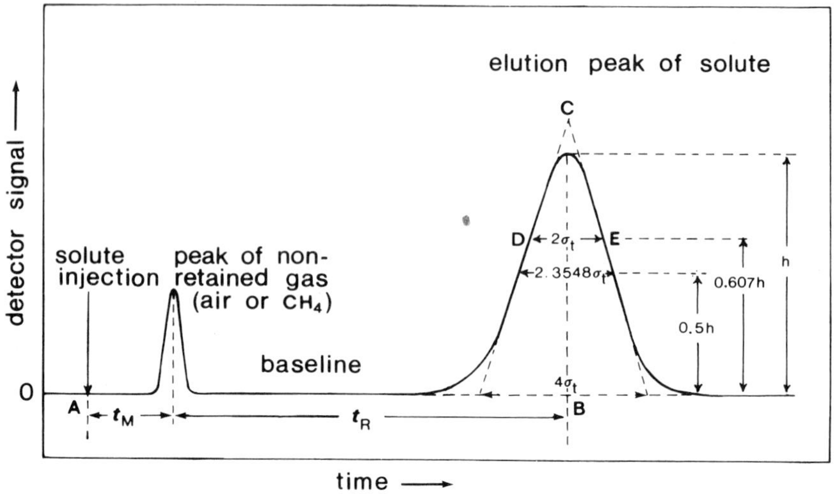

Fig. 1.5 Typical appearance of an elution chromatogram after the injection of a single solute.

$$V_N = V_R^o - V_M^o = \dot{V}t_R - \dot{V}t_M = \dot{V}(t_R - t_M) \qquad (1\text{-}23)$$

From V_N, the partition coefficient K or β is easily calculated using Eqs. (1-17) or (1-18), respectively.

As it is shown in Fig. 1.5, the time, t_M, is found from the position of the peak due to the air, which is always introduced with the sample. This is recorded by a thermal conductivity detector, but not by a flame ionization detector. In the latter case, the introduction of another inflammable and nonretained gas, like CH_4, is required.

Another quantity frequently quoted in gas chromatographic studies is the <u>specific retention volume</u>, V_g, which is characteristic of a particular solute, stationary phase, and carrier gas. It is the net retention volume at 0°C for unit weight of stationary phase, and is thus calculated by the relation.:

$$V_g = \frac{273.15V_N}{T_c W} \qquad (1\text{-}24)$$

C. Broadening Factors in Gas Chromatography

The elution curve of a solute component (cf. Fig. 1.5) is a plot of the concentration, c_G, of this component at the end of the chromatographic column ($x = l$) versus time, t. This is because the chromatographic signal of the detector is proportional to C_G in the exit gas. Theoretically the function $c_G = c_G(t)$ is given by Eq. (1-10) with $x = l$, which predicts a very sharp and narrow peak at $t = (1 + k)l/v + t_R$ (cf. Eqs. (1-11) and (1-14)). The peak is supposed to be so sharp that for t a little different from t_R the concentration c_G is zero. This is due to the fact that the dependence of c_G on t is described by a delta function $\delta(t - a)$, as Eq. (1-10) shows. In spite of that, the elution curve of Fig. 1.5 is broadened compared to that predicted by the theory, and this is due to various so-called broadening factors. The most important of these are related to nonfulfilment of the assumptions under which Eq. (1-10) was derived, i.e.,

1. Nonlinearity of the isotherm
2. Nonnegligible axial diffusion of the gas in the chromatographic column
3. Unsharp input distribution of the vapor
4. Noninstantaneous equilibration of the solute components between the mobile and the stationary phases; i.e., nonideal chromatography.

If the elution curve $c_G = c_G(t)$ is taken as a continuous statistical distribution in t, the broadening of the curve is proportional to the standard deviation, σ_t, of the distribution, as measured by the width of the peak at varius heights from the baseline (cf. Fig. 1.5). If the shape of the peak is gaussian, the width at the inflection points D, E, located at 0.607 h, equals $2\sigma_t$. At half of the maximum height the width is equal to $2\sqrt{2\ln 2}\,\sigma_t = 2.3548\sigma_t$.

Each of the broadening factors produces a variance σ_i^2 on an instantaneously introduced zone of sample. The total variance is equal to the sum of the variances due to all factors:

$$\sigma_{total}^2 = \sum_i \sigma_i^2 \tag{1-25}$$

The retention time, t_R, represents the mean value of the distribution of t, and therefore it is given by the abscissa of the center of gravity of the elution peak. This is approximately the same with the abscissa of point C in Fig. 1.5.

Let us now examine in more detail some of the broadening factors.

1. Longitudinal Diffusion in the Gas Phase

The term longitudinal diffusion is used to denote axial diffusion of the gas in the chromatographic column. This includes true longitudinal molecular diffusion and apparent or eddy diffusion. The first occurs because of concentration gradients within the carrier gas along the column, but eddy diffusion results from uneven velocity profiles because of zigzag paths of unequal lengths and widths. In what follows, we assume only true longitudinal diffusion, obeying the one dimensional diffusion equation (Fick's second law):

$$\frac{\partial c_G}{\partial t} = D_G \frac{\partial^2 c_G}{\partial y^2} \tag{1-26}$$

where c_G is the gas phase solute concentration, D_G its diffusion coefficient in that phase (carrier gas), and y the length coordinate along the column axis similar to x of Fig. 1.4, but with its zero at the peak maximum in the center of the vapor zone. We assume that initially the zone is a very narrow one as that described by a delta function, so that the initial condition at $t = 0$ is

$$c_G(0,y) = \frac{m}{a_G}\ \delta(y) \tag{1-27}$$

with m denoting as before the total amount of solute in mol, and a_G the volume of gas phase per unit length of column (cm^2). The diffusion of the above narrow band occurs whether the zone is stationary or being eluted through the column. To make calculations simpler, we ignore at the moment the movement of the band along the column, i.e., the chromatographic process itself, assuming that the elution of the vapor does not affect its diffusion. Therefore, we only seek a solution of the partial differential equation (1-26) under the initial condition (1-27), and the boundary conditions: 1. that c_G remains finite as $y \to \infty$, 2. that at $y = 0$ we have the maximum value of c_G, i.e.,

$$\left(\frac{\partial c_G}{\partial y}\right)_{y=0} = 0 \tag{1-28}$$

The solution can most easily be found by using Laplace transforms [14], as was done before for solving Eq. (1-4). Transformation with respect to time (transform parameter p) of Eq. (1-26) gives

$$pC_G - \frac{m}{a_G}\ \delta(y) = D_G \frac{d^2C_G}{dy^2} \tag{1-29}$$

the capital letter C_G denoting the t transformed function of c_G. This equation, written in the form

$$\frac{d^2C_G}{dy^2} - q^2C_G = -\frac{m}{a_G D_G}\,\delta(y) \tag{1-30}$$

where

$$q^2 = \frac{p}{D_G} \tag{1-31}$$

is an ordinary linear second-order differential equation. It can be integrated by further using Laplace transformation with respect to y (transform parameter s):

$$s^2\bar{C}_G - sC_g(0) - C'_G(0) - q^2\bar{C}_G = -\frac{m}{a_G D_G}$$

or

$$\bar{C}_G = C_G(0)\frac{s}{s^2 - q^2} + \frac{C'_G(0)}{s^2 - q^2} - \frac{m}{a_G D_G} \cdot \frac{1}{s^2 - q^2} \tag{1-32}$$

where $\bar{C}_G$ is the double Laplace transform of c_G with respect to t and y, $C_G(0)$ is the t transform of c_G at y = 0, and $C'_G(0)$ the y derivative of C_G at y = 0; i.e., $(dC_G/dy)_{y=0}$. Inverse Laplace transformation of Eq. (1-32) with respect to the parameter s gives (15):

$$C_G = C_G(0)\cosh qy + \frac{C'_G(0)}{q} \sinh qy - \frac{m}{a_G D_G q} \sinh qy \tag{1-33}$$

In this solution $C_G(0)$ and $C'_G(0)$ are in essence the two constants of integration of Eq. (1-30) and their values can be found by using the boundary conditions mentioned earlier. First, $C'_G(0) = 0$ because of Eq. (1-28). Second, we write (1-33) in the form

$$C_G = \left[C_G(0) - \frac{m}{a_G D_G q}\right]\frac{\exp(qy)}{2} + \left[C_G(0) + \frac{m}{a_G D_G q}\right]\frac{\exp(-qy)}{2} \tag{1-34}$$

and choose the value zero for the first bracket [ ] on the right-hand side, so that C_G remains finite as $y \to \infty$. This gives the value of $C_G(0)$ as

$$C_G(0) = \frac{m}{a_G D_G q} \qquad (1\text{-}35)$$

and by substituting it into the second bracket of (1-34), we obtain

$$C_G = \frac{m}{a_G D_G q} \exp(-\, qy) \qquad (1\text{-}36)$$

By taking now the inverse transform of this equation with respect to p, we have

$$c_G = \frac{m}{a_G} \cdot \frac{1}{(\pi D_G t)^{1/2}} \exp(-\, y^2/4D_G t) \qquad (1\text{-}37)$$

To normalize this function to unity, it is necessary that

$$\mathscr{L}_{-\infty}^{\infty}\; c_G dy = 2\mathscr{L}_{-\infty}^{\infty}\; c_G dy = 1 \qquad (1\text{-}38)$$

Instead of calculating the integral using the analytical expression of c_G from Eq. (1-37), we can take advantage of a property of the Laplace transforms, i.e.,

$$\mathscr{L}_{-\infty}^{\infty}\; C_G dy = \lim_{s \to 0} \bar{C}_G = \frac{m}{a_G D_G q^2} = \frac{m}{a_G p} \qquad (1\text{-}39)$$

In place of $\bar{C}_G$ we have used the right-hand side of Eq. (1-32), setting $C'_G(0)$ because of (1-28). Also, q^2 was substituted by p/D_G according to the definition (1-31). Now, the p inverse transform of (1-39) gives

$$\mathscr{L}_{-\infty}^{\infty}\; c_G dy = m/a_G$$

and this, combined with (1-38), leads to the normalization condition as $m/a_G = 1/2$. Substituting 1/2 for m/a_G in Eq. (1-37), we have

$$c_G = \frac{1}{2(\pi D_G t)^{1/2}} \exp\left(-\frac{y^2}{4D_G t}\right) \tag{1-40}$$

Comparison with the probability density function of the normal or gaussian distribution:

$$\Phi(y) = \frac{1}{\sigma(2\pi)^{1/2}} \exp\left[-\frac{(y-\bar{y})^2}{2\sigma^2}\right] \tag{1-41}$$

shows that c_G is a gaussian function of y with mean $\bar{y} = 0$ and vaiance

$$\sigma_x^2 = 2D_G t \tag{1-42}$$

This relation, known as Einstein's law of diffusion, together with Eq. (1-40) explains why the chromatographic column is a gaussian operator and the solute vapor zone, introduced as a delta function, will come out at the other end of the column as a gaussian function had longitudinal diffusion been the only broadening factor.

The variance in Eq. (1-42) and the respective standard deviation, σ_x, are expressed in length units along the column coordinate. To transform it in time units it is necessary to multiply it by $(1 + k)/v$, as Eq. (1-11) shows:

$$\sigma_t = \sigma_x \frac{1+k}{v} = \sigma_x \frac{(1+k)a_G}{\dot{V}} \tag{1-43}$$

Some properties of the normal distribution curve, described by Eq. (1-41), are worth mentioning, since they also apply to the chromatographic peaks. First, the curve is symmetrical about the line y = 0 through the mean of the distribution. Further, the ordinates decrease rapidly as $|y|$ increases. Lastly, by equating to zero the second derivative of c_G in Eq. (1-40), one can easily verify that the points of inflection on the curve are given by $y = \pm\sigma_x$. This is the reason that the width at the inflection points D, E in Fig. 1.5 is equal to $2\sigma_t$.

2. Noninstantaneous Equilibration of Solute Between the Two Phases

If the rate of exchange of solute between the mobile and the stationary phase is not infinite, equilibrium between phases will not be established instantaneously. This effect is often called resistance to mass transfer.

The mass balance equation is again Eq. (1-2), but the isotherm (1-3) cannot be substituted into it, as was done there, since equilibration between phases is slow. If equilibrium is reached according to a first-order law, the rate of change of the solute concentration in the stationary phase is

$$\frac{\partial c_S}{\partial t} = k_1(c_S^* - c_S) = k_1(Kc_G - c_S) \tag{1-44}$$

where k_1 is a rate constant and K the partition coefficient of the solute between the two phases. Thus Kc_G is the concentration c_S^* in the stationary phase which would be in equilibrium with the actual concentration in the gas phase.

Substituting the right-hand side of (1-44) for $\partial c_S/\partial t$ in Eq. (1-2), one obtains

$$\frac{\partial c_G}{\partial t} + rk_1(Kc_G - c_S) + v\frac{\partial c_G}{\partial x} = 0 \tag{1-45}$$

where $r = a_S/a_G$ (volume ratio of the two phases). The method of Laplace transforms is again used to solve the system of partial differential equations (1-44) and (1-45), under the same initial and boundary conditions as before, i.e., Eqs. (1-6) and

(1-7). The transformation with respect to time (parameter p) of (1-44) and (1-45) leads to

$$pC_S = k_1KC_G - k_1C_S$$

$$pC_G + rk_1KC_G - rk_1C_S + v\frac{dC_G}{dx} = 0$$

Elimination of C_S is effected by solving the first of these equations for C_S and then substituting it into the second. The result is

$$\frac{dC_G}{dx} + \frac{\Omega}{v} C_G = 0 \tag{1-46}$$

where

$$\Omega = p + rk_1K - \frac{rk_1{}^2K}{p + k_1} \tag{1-47}$$

Separation of variables and integration of (1-46) with respect to x, using the t transformed boundary condition (1-7) to evaluate the constant of integration, gives

$$C_G = \frac{m}{\dot{V}} \exp\left(- \Omega \frac{x}{v}\right) \tag{1-48}$$

where $\dot{V} = a_G v$. Now, inverse Laplace transformation of this equation with respect to p gives [15]

$$c_G = \frac{m}{\dot{V}} \left\{ \exp\left(\frac{-rk_1Kx}{v}\right) \cdot \exp\left[-k_1\left(t - \frac{x}{v}\right)\right] \right.$$
$$\cdot \left(\frac{rk_1^2Kx/v}{t - x/v}\right)^{1/2} I_1\left[2\left(\frac{rk_1^2Kx}{v}\right)^{1/2}\left(t - \frac{x}{v}\right)^{1/2}\right] u\left(t - \frac{x}{v}\right)$$
$$\left. + \exp\left(\frac{- rk_1Kx}{v}\right) \cdot \delta\left(t - \frac{x}{v}\right)\right\} \tag{1-49}$$

where I_1 is the hyperbolic Bessel function of first order, with the argument enclosed in the brackets [ ], and $u(t - x/v)$ the Heaviside unit step function, which equals 0 for $t < x/v$ and 1 for $t \geq x/v$.

At the detector, i.e., at x = l (cf. Fig. 1-4), the various groups of variables of Eq. (1-49) become

$$\frac{rk_1Kx}{v} = \frac{k_1kl}{v} = k_1kt_M = k_1(t_R - t_M)$$

$$\frac{rk_1^2Kx}{v} = k_1^2(t_R - t_M)$$

$$t - \frac{x}{v} = t - t_M$$

Substituting these relations into Eq. (1-49) and observing that for $t > t_M$, i.e., after a short time from the injection of the solute, $u(t - t_M)$ becomes unit, whereas $\delta(t - t_M)$ becomes zero, we are left with the following expression at the detector:

$$c_G = \frac{m}{\dot{V}} \exp\left[- k_1(t_R - t_M + t - t_M)\right] \cdot \frac{k_1(t_R - t_M)^{1/2}}{(t - t_M)^{1/2}} I_1[2k_1(t_R - t_M)^{1/2}(t - t_M)^{1/2}] \quad (1\text{-}50)$$

Adopting the approximation for the hyperbolic Bessel functions

$$I_n(z) \simeq e^z/(2\pi z)^{1/2}$$

which is valid for $z > 10$, Eq. (1-50) becomes

$$c_G = \frac{m}{\dot{V}} \left(\frac{k_1}{4\pi}\right)^{1/2} \frac{(t_R - t_M)^{1/4}}{(t - t_M)^{3/4}} \exp[- k_1(\sqrt{t_R - t_M} - \sqrt{t - t_M})^2] \quad (1\text{-}51)$$

The expresion $(\sqrt{t_R - t_M} - \sqrt{t - t_M})^2$ in the exponent can be written

$$(\sqrt{t_R - t_M} - \sqrt{t - t_R + t_R - t_M})^2 = (\sqrt{a} - \sqrt{z + a})^2$$

For small values of $z = t - t_R$, i.e., for times near the peak maximum, this expression can be expanded in a McLaurin series in z, retaining only the first non-zero term, which is $z^2/4a$. With this approximation, and by setting $t_R \simeq t$ in the preexponential factor and the denominator of the exponent, Eq. (1-51) finally becomes

$$c_G = \frac{m}{\dot{V}} \left[\frac{k_1}{4\pi(t - t_M)}\right]^{1/2} \exp\left[-\frac{k_1(t - t_R)^2}{4(t - t_M)}\right] \tag{1-52}$$

The normalization condition for this function of t is found by integrating it between the limits t = 0 and t = ∞, or by finding the limit of (1-48) for p → 0 and equating it to 1. The result is $m/\dot{V} = 1$.

Comparing now Eq. (1-52) with Eq. (1-41), we conclude that it is a gaussian function of t with mean $\bar{t} = t_R$ and variance

$$\sigma_t^2 = \frac{2(t - t_M)}{k_1} \tag{1-53}$$

To transform the variance in length units we make use of Eq. (1-11):

$$\sigma_x^2 = \frac{2kvx}{k_1(1 + k)^2} \tag{1-54}$$

In conclusion, Eq. (1-52) shows that a finite rate and non-instantaneous equilibration of a solute component between the mobile and the stationary phases broadens the delta function input distribution into an approximately normal or gaussian distribution, having the same mean with that in the absence of the broadening factor.

There remains to specify the nature of the rate constant, k_1, for the transfer of solute between the stationary and the mobile phase. There are several mechanisms to explain the non-instantaneous equilibration of solute, the most common of which seems to be the slow diffusion of the solute vapor in the stationary phase. If the diffusion coefficient in this phase is D_S and the depth of the phase d (thickness of the liquid layer on solid support in GL chromatography or depth of pores of uniform bore in GS chromatography), the mass transfer coefficient in this phase is D_S/d. The rate constant, k_1, is proportional to this and to the effective area per unit volume of the phase, $A/V = A/Ad = 1/d$:

$$k_1 = \frac{\pi^2}{4} \cdot \frac{D_S}{d} \cdot \frac{1}{d} = \frac{\pi^2 D_S}{4d^2} \qquad (1\text{-}55)$$

The constant of proportionality, $\pi^2/4$, has been derived by van Deemter et al. (16). Substituting this into Eq. (1-54), one obtains

$$\sigma_x^2 = \frac{8}{\pi^2} \cdot \frac{d^2 v}{D_S} \cdot \frac{kx}{(1 + k)^2} \qquad (1\text{-}56)$$

D. Theoretical Plates and van Deemter Equation

The analogy between fractional distillation and chromatography leads to the definition of theoretical plates, N, for the chromatographic column by the relation:

$$N = \frac{t_R^2}{\sigma_t^2} = \frac{l^2}{\sigma_x^2} \qquad (1\text{-}57)$$

The apparent plate height $\hat{H}$ is given by

$$\hat{H} = \frac{1}{N} = \frac{\sigma_x^2}{l} \tag{1-58}$$

Apparently the separation efficiency of the column depends on the total σ^2, as given by Eq. (1-25), and this in turn on the various broadening factors. For packed columns with relatively large volumes of stationary phase, the major contributions to peak broadening are those described in detail in the previous paragraph; i.e., longitudinal diffusion in the gas phase [Eq. (1-42)] and noninstantaneous equilibration of solute between the two phases owing to slow diffusion in the stationary phase (Eq. [1-56]). According to Eq. (1-25), the total broadening is measured by the total variance, which is obtained by adding Eqs. (1-42) and (1-56), after some modifications:

$$\sigma_{total}^2 = \frac{2D_G t_R \gamma}{1 + k} + \frac{8}{\pi^2} \cdot \frac{d^2 v}{D_S} \cdot \frac{kl}{(1 + k)^2} \tag{1-59}$$

The first term was multiplied by the so-called <u>tortuosity</u>, γ, to account for the fact that diffusion is not exactly along the column axis, but along the tortuous paths between the particles of the column packing. It was also divided by 1 + k because the fraction of the solute in the vapor phase is 1/(1 + k). Finally, both terms are specified for the total column time, i.e., t_R, and the total column length, l. If now $t_R/(1 + k)$ is replaced by l/v according to Eq. (1-14), and the right-hand side of Eq. (1-59) is substituted for σ_x^2 in Eq. (1-58), the plate height is obtained as a function of the linear velocity of the carrier gas:

$$\hat{H} = A + \frac{2\gamma D_G}{v} + \frac{8}{\pi^2} \cdot \frac{d^2}{D_S} \cdot \frac{k}{(1 + k)^2} v \tag{1-60}$$

The term A is added to account for flow-independent contributions to $\hat{H}$. Eq. (1-60), called the van Deemter equation

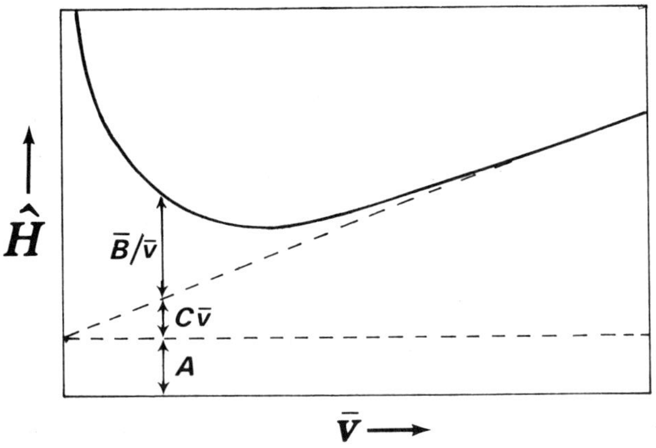

Fig. 1.6 Plot of van Deemter equation (1-61) showing the contributin of the three terms to $\hat{H}$ at a specified mean linear velocity $\bar{v}_1$.

[16], is classically written

$$\hat{H} = A + \frac{\bar{B}}{\bar{v}} + C\bar{v} \tag{1-61}$$

where $\bar{B} = 2\gamma D_G$ and C is given by the coefficient of v in the third term of Eq. (1-60). Giddings (11) has developed a generalized nonequilibrium theory and tabulated a great number of C terms owing to various mechanisms.

A plot of the van Deemter equation is shown in Fig. 1.6. It is clear from this plot (an hyperbola) that the optimum performance of the column corresponds to the minimum of the curve, occurring at $\bar{v} = (\bar{B}/C)^{1/2}$. The minimum plate height is thus given by

$$\hat{H} = A + 2(\bar{B}C)^{1/2} \tag{1-62}$$

V. PHYSICOCHEMICAL MEASUREMENTS BY GAS CHROMATOGRAPHY

In addition to chemical analysis, gas chromatography offers many possibilities for physicochemical measurements. Some of these methods lead to very precise and accurate results with relatively cheap instrumentation and a very simple experimental setup. They are widely used today, a fact emphasized by the editions of two books (17,18) dealing only with such physicochemical measurements. All are based on the traditional techniques breifly described in section III of this Chapter; i.e., elution development, frontal analysis, and displacement development, under constant gas flowrate.

Among the first physicochemical measurements are adsorption studies relating to determination of adsorption isotherms and thermodynamic parameters for adsorption. The details of this application must be found elsewhere (19). Here it is worth mentioning that the simple observation of a chromatographic elution peak gives qualitative information about the shape of the adsorption isotherm, as depicted in Fig. 1.7. The peaks are symmetrical only when the isotherm is linear, whereas a convex or a concave isotherm produces peak asymmetry either in the back or in the front of the peak profile.

Some recent advances on determination of gas-solid adsorption isotherms by the so-called step and pulse method have been made by Guiochon and his coworkers (20,21). Also, Jaulmes et al. (22–24) have made a thorough study of peak profiles in nonlinear gas chromatography. They proposed a theoretical model for the profile of elution peaks accounting both for the influence of the isotherm curvature at zero concentration and for the perturbation of the flowrate due to solute exchange between the mobile and the stationary phase. The physical reliability of the theoretical model has been demonstrated by the fact that the parameters of the adsorption isotherm deter-

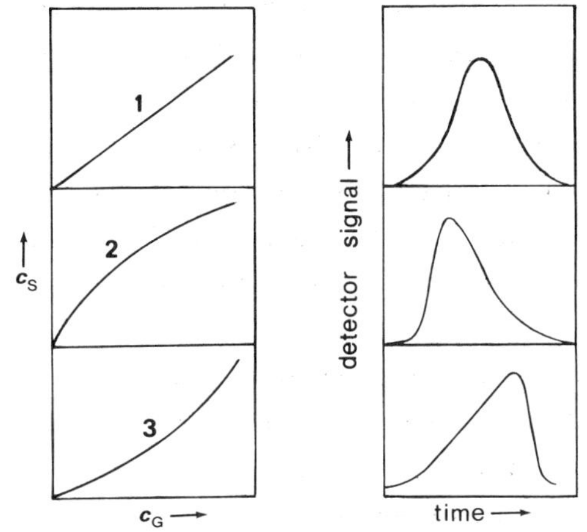

Fig. 1.7 Three isotherm forms (1, linear; 2, convex; 3, concave), and the corresponding peak shapes (on the right).

mined from their experimental data are in good agreement with the results obtained by independent techniques of isotherm determination. The model has been further applied for description of the experimental elution profiles observed in gas chromatography with overloaded capillary columns (25).

Heats and entropies of adsorption are determined by GS chromatography using the equation (26)

$$\ln V_N = \ln(RTn_s) + \frac{\Delta S}{R} - \frac{\Delta H}{R} \cdot \frac{1}{T} \qquad (1\text{-}63)$$

wher n_s(mol) is the total amount of the solute in the adsorbed state, ΔH and ΔS being the differential enthalpy and differential entropy of adsorption, respectively. By determining the net retention volume, V_N, at various temperatures, we can calculate ΔH from the slope and ΔS from the intercept of the plot of $\ln V_N$ against $1/T$.

The so-called thermodynamic compensation effect, i.e., a linear dependence of ΔS on ΔH, often observed in such studies, can be explained by expressing V_N in terms of molecular properties with the help of statistical mechanics (26). For this purpose, a heterogeneous surface on the solid adsorbent and an exponential distribution of adsorption sites with respect to the energy are assumed. From the results of adsorption of alkanes and cycloalkanes on aluminum oxide, the frequency of the vibrations of the adsorbate molecules normal to the surface was calculated.

In a further adsorption study by GS chromatography based on the compensation effect (27), two new parameters were extracted from the experimental data. These parameters are considered to be more fundamental quantities characterizing the adsorption phenomenon than the enthalpy and entropy of adsorption .

As regards the condition for a compensation effect to be observed, this is found as follows (27). It is a rather accepted view that for the existence of a linear dependence of ΔS on ΔH it is necessary that the straight lines predicted by Eq. (1-63) intersect at about the same point. This is because at the point of intersection (defined by the pair of values $\ln V_{N_s}$, $1/T_s$) Eq. (1-63) reads (neglecting the first insignificant term):

$$\ln V_{N_s} \simeq \frac{\Delta S_i}{R} - \frac{\Delta H_i}{R} \cdot \frac{1}{T_s} \tag{1-64}$$

where ΔS_i and ΔH_i refer to values calculated from line i. On rearrangement one obtains

$$\Delta S_i \simeq R \ln V_{N_s} + \frac{1}{T_s} \cdot \Delta H_i \tag{1-65}$$

i.e., a linear dependence of ΔS_i on ΔH_i. However, it is a rather improbable situation that all lines intersect at about the same point, and the frequency of occurrence of a compensation effect seems to suggest that this is not the only possible situation. In the analysis of some results for example, it is not at all apparent from the plots of Eq. (1-63) that a linear relationship between ΔS and ΔH should exist for hydrocarbons with the same number of carbon atoms and an alumina surface covered with different salts. The explanation that such linear relationships exist lies in the following. Take the intersection of line i with line j. For this point Eq. (1-64) is

$$\ln V_{N_{i,j}} = \frac{\Delta S_i}{R} - \frac{\Delta H_i}{R} \cdot \frac{1}{T_{i,j}} \tag{1-66}$$

By summing over j for all $j \neq i$, we obtain

$$\sum_{j=1}^{n} \ln V_{N_{i,j}} = n \frac{\Delta S_i}{R} - \frac{\Delta H_i}{R} \sum_{j=1}^{n} \frac{1}{T_{i,j}}$$

or, on division by n, and rearrangement

$$\Delta S_i = R < \ln V_{N_{i,j}} > + < \frac{1}{T_{i,j}} > \Delta H_i \tag{1-67}$$

where $< \ln V_{N_{i,j}} >$ and $< 1/T_{i,j} >$ are mean values defined by $(1/n)\ \Sigma_j \ln V_{N_{i,j}}$ and $(1/n)\Sigma_j(1/T_{i,j})$, respectively. Thus, for a linear dependence of ΔS_i on ΔH_i to exist, it is only necessary that the <u>mean values</u> of the coordinates of the intersection points of each line with all the rest are about the same.

The application of the above conclusion can be demonstrated for the case of pentane and cyclopentane adsorbed on alumina modified with 20% w/w of LiCl, NaCl, RbCl, CsCl, NaF, NaBr, and NaI (27). There are 14 plots of Eq. (1-63) and 91 points of intersection of these lines. The ordinates of these points range from −50.9 to 269.0, and the abscissae from -13.7×10^{-3}

to 82.4×10^{-3}, whereas the mean values of ordinates and abscissae lie between 2.5 and 6.2 and between 2.3×10^{-3} and 3.6×10^{-3}, respectively. The ranges of the mean values are narrow enough to create a compensation effect according to Eq. (1-67).

A series of adsorption studies of various hydrocarbons on graphitized carbon black and graphite has been published by Vidal-Madjar et al. (28–32). In these studies, retention volumes and thermodynamic functions of adsorption are calculated with the help of molecular statistical theory, and compared with those found experimentally by GS chromatography. A good agreement between predicted and measured values was observed.

Before these recent advances, a wealth of studies had been conducted on physical properties of pure substances, solution thermodynamics, formation of complexes, phase transitions, virial coefficients, diffusion processes, and chemical kinetics of reactions taking place on the chromatographic column. All these physicochemical measurements are described in great detail in the two books (17,18) mentioned earlier in this section, so that their repetition here does not seem in order.

VI. THE CARRIER GAS FLOWRATE PERTURBATION

Although there would be no gas chromatography without a mobile gas phase, i.e., a carrier gas, its linear velocity, v, or volume flowrate, $\dot{V} = a_G v$, remains constant throughout a single experiment in most gas chromatographic studies, or analytical applications. Thus, this magnitude is usually treated as an adjustable parameter of gas chromatographic equations. Following, however, the widespread use of temperature programming in gas chromatographic analysis, the programming of the carrier gas inlet pressure, and hence its flowrate, also had been reported and reviewed (33,34). In spite of the development of various programming modes (e.g., step programming, continuous linear and nonlinear programming), and the existence of commercial units permitting the general use of the technique, flow (pressure) programming has not been used to extract information of a physicochemical nature in gas chromatography. Its uses have been limited to analytical applications.

Except flow programming, there are two other kinds of flowrate perturbations imposed on the carrier gas. These are the stopped-flow and reversed-flow techniques. Both are very simple to apply and consist of either stopping the carrier gas flow for short time intervals, or reversing the direction of its flow from time-to-time. Experimentally, this is most easily done by using shut-off valves in the first technique and a four- or six-port gas sampling valve in the second. Thus, sophisticated mechanical, pneumatic, or other special systems are not required as in flow programming gas chromatography.

To the best of our knowledge, the first who used the stopping of the carrier gas flow for varying time periods were Knox and McLaren (35), with the purpose of producing extra broadening of the chromatographic peaks for measuring gas diffusion coefficients. However, the stopped-flow method was substantially introduced in 1967 by Phillips and his coworkers (36), not for a mere broadening of an existing chromatographic peak, but to create new very narrow and symmetrical peaks on an asymmetrical elution curve due to a chemical reaction on the chromatographic column. The reversed-flow technique was introduced in its preliminary form in 1980 (37).

Both these techniques have solely been used for physicochemical measurements and constitute the object of this book. In the next chapter, the stopped-flow technique and its applications are exposed, whereas the rest of the book is devoted to the reversed-flow technique.

LIST OF SYMBOLS

a_G	Volume of gas phase per unit length of column, or cross-sectional area of void space
a_S	Volume of stationary phase per unit length of column
$\bar{B}$	Coefficient in the van Deemter equation (1-60)
C	Coefficient in the van Deemter equation (1-60)
c_G	Concentration of a solute vapor in the gas phase of the column
C_G	Laplace transform of c_G with respect to t
$\bar{C}_G$	Double Laplace transform of c_G with respect to t and y
c_S	Concentration of solute in the stationary phase
c_S^*	Concentration in the stationary phase in equilibrium with the actual concentration in the gas phase
C_S	Laplace transform of c_S with respect to t
d	Thickness of the liquid layer on solid support or depth of pores of uniform bore
D_G	Gas diffusion coefficient
D_S	Diffusion coefficient in the stationary phase
$\hat{H}$	Apparent plate height
I_n	Hyperbolic Bessel function of order n
j	Compressibility factor defined by Eq. (1-20)
k	Partition ratio
k_1	Rate constant
K	Partition coefficient or distribution constant
l	Total column length
m	Total amount of the component to be analyzed
m_S	Mass of stationary phase per unit length of column
N	Number of theoretical Plates
p	Transform parameter with respect to t
p_i	Inlet pressure of column
p_0	Pressure at the flowmeter
p_w	Vapor pressure of water at the flowmeter temperature
q	Parameter defined by Eq. (1-31)

r	Volume ratio of stationary to mobile phases (= a_S/a_G
R_f	Retardation factor defined by Eq. (1-13)
s	Transform parameter with respect to y
t	Time variable
t_M	Hold-up time, or dead time
t_R	Retention time defined by Eq. (1-14)
T_c	Column temperature
T_f	Flowmeter temperature
u	Linear velocity of solute component defined by Eq. (1-12)
W	Total weight of stationary phase
x,y	Distance coordinates along the column
v	Linear velocity of carrier gas in interparticle space of the column
v_c	Chart speed
V	Volume of carrier gas passed through the column
$\dot{V}$	Volume flowrate of carrier gas
$\dot{V}_c$	Volume flowrate at column temperature given by Eq. (1-19)
$\dot{V}_f$	Volume flowrate measured at the flowmeter temperature
V_g	Specific retention volume defined by Eq. (1-24)
V_M^o	Corrected hold-up volume, or dead volume
V_N	Net retention volume
V_R^o	Corrected retention volume
V_S	Total volume of stationary phase
z_M	Distance on recorder chart corresponding to t_M
z_R	Distance on recorder chart corresponding to t_R
β	Partition coefficient in $cm^3 g^{-1}$
γ	Tortuosity
σ_i^2	Variance of the peak distribution due to factor i
σ_{total}^2	Total variance of the peak distribution
σ_t	Standard deviation in time units

σ_x Standard deviation in length units

Ω Functions defined by Eq. (1-47)

REFERENCES

1. M. Tswett, Ber. Dtsch. Bot. Ges., 24:316, 384 (1906).
2. R. Kuhn and E. Lederer, Chem. Ber., 64:1349 (1931).
3. R. Kuhn, A. Winterstein, and E. Lederer, Hoppe Seylers Z. Physiol. Chem., 197:141 (1931).
4. J. N. Wilson, J. Am. Chem. Soc., 62:1583 (1940).
5. E. Glueckauf, J. Chem. Soc., 1302 (1947).
6. A. T. James and A. J. P. Martin, Biochem. J., 50:679 (1952).
7. L. S. Ettre and A. Zlatkis (eds.), 75 Years of Chromatography—A Historical Dialogue, Elsevier, Amsterdam, 1979.
8. A. B. Littlewood, Gas Chromatography, 2nd ed., Academic Press, New York, 1970.
9. H. Purnell, Gas Chromatography, Wiley, 1967.
10. R. L. Grob, Modern Practice of Gas Chromatography, Wiley-Interscience, New York, 1977.
11. J. C. Giddings, Dynamics of Chromatography, Dekker, New York, 1965.
12. J. C. Sternberg, Adv. Chromatogr., 2:205 (1966).
13. D. DeVault, J. Am. Chem. Soc., 65:532 (1943).
14. M. Boas, Mathematical Methods in the Physical Sciences, Wiley, New York, 1966.
15. F. Oberhettinger and L. Badii, Tables of Laplace Transforms, Springer-Verlag, Berlin, 1973.
16. J. J. van Deemter, F. J. Zuiderweg, and A. Klinkenberg, Chem. Sci., 5:271 (1956).
17. R. J. Laub and R. L. Pecsok, Physicochemical Applications of Gas Chromatography, Wiley, New York, 1978.

18. J. R. Conder and C. L. Young, Physicochemical Measurement by Gas Chromatography, Wiley, Chichester, 1979.

19. A. V. Kiselev and Ya. I. Yashin, Gas Adsorption Chromatography, Plenum Press, New York, 1969.

20. F. Dondi, M. F. Gonnord, and G. Guiochon, J. Colloid Interface Sci., 62:303 (1977).

21. F. Dondi, M. F. Gonnord, and G. Guiochon, J. Colloid Interface Sci., 62:316 (1977).

22. A. Jaulmes, C. Vidal-Madjar, A. Ladurelli, and G. Guiochon, J. Phys. Chem., 88:5379 (1984).

23. A. Jaulmes, C. Vidal-Madjar, M. Gaspar, and G. Guiochon, J. Phys. Chem., 88:5385 (1984).

24. A. Jaulmes, C. Vidal-Madjar, H. Colin, and G. Guiochon, J. Phys. Chem., 90:207 (1986).

25. P. Cardot, I. Ignatiadis, A. Jaulmes, C. Vidal-Madjar, and G. Guiochon, J. High Resolution Chromatogr. Chromatogr. Commun., 8:591 (1985).

26. N. A. Katsanos, A. Lycourghiotis, and A. Tsiatosios, J. Chem. Soc., Faraday Trans. I, 74:575 (1978).

27. G. Karaiskakis, A. Lycourghiotis, and N. A. Katsanos, Z. Physik. Chem. (N.F.), 111:207 (1978).

28. M. F. Gonnord, C. Vidal-Madjar, and G. Guiochon, J. Chromatogr. Sci., 12:839 (1974).

29. C. Vidal-Madjar, M. F. Gonnord, M. Goedert, and G. Guiochon, J. Phys. Chem., 79:732 (1975).

30. C. Vidal-Madjar, M. F. Gonnord, and G. Guiochon, J. Colloid Interface Sci., 52:102 (1975).

31. C. Vidal-Madjar, G. Guiochon, and F. Dondi, J. Chromatogr., 291:1 (1984).

32. C. Vidal-Madjar and E. Bekassy-Molnar, J. Phys. Chem., 88:232 (1984).

33. R. P. W. Scott, Progress in Gas Chromatography J. H. Purnell, ed.), Interscience, 6:271 (1968).

34. L. S. Ettre, L. Májor, and J. Takács, Adv. Chromatogr., 8:271 (1969).

35. J. H. Knox and L. McLaren, Anal. Chem., 36:1477 (1964).

36. C. S. G. Phillips, A. J. Hart-Davis, R. G. L. Saul, and J. Wormald, J. Gas Chromatogr., 5:424 (1967).

37. N. A. Katsanos and I. Georgiadou, J. Chem. Soc., Chem. Commun., 242 (1980).

2

The Stopped-Flow Technique

I. Introduction

II. Experimental Setup

III. Ideal Stopped–Flow Gas Chromatography with Chemical Reactions in the Stationary Phase

A. General theoretical analysis
B. Nonopposing reaction on one kind of active site with no intermediate
C. Nonopposing reactions on three kinds of active sites
D. Nonopposing reactions on three kinds of active sites
E. Catalytic conversions calculated by stopped-flow gas chromatography

IV. Nonideal Stopped–Flow Gas Chromatography

V. Measurement of Diffusion Coefficients in Gases

List of Symbols

References

I. INTRODUCTION

The stopped-flow technique of Phillips and his co-workers (1) was introduced in 1967 for measuring rate constants of simple surface-catalyzed reactions. It is well known that complicated experimental arrangements are usually required to carry out kinetic experiments for such reactions. Moreover, the kinetic parameters so determined are usually apparent, and not true ones. The stopped-flow technique requires a very simple experimental set-up; i.e., a slightly modified gas chromatograph, and leads to the determination of <u>true</u> kinetic parameters. It is an accurate and easy method for studying heterogeneous catalytic reactions, and can be carried out even by students having the usual laboratory experience.

Suppose that a substance A, capable of undergoing a simple first-order reaction

$$A \xrightarrow{k_1} B + C \tag{2-1}$$

is injected into a gas chromatographic column. If the column is filled with a solid, which acts both as a catalyst for the reaction and as separating phase of product B from the reactant A, the chromatographic trace on the recorder will look somewhat like Fig. 2.1. In this figure, it is assumed that the product B has a smaller retention time than the reactant A, and C is not detected (e.g., H_2O with a flame ionization detector).

The rate of the reaction, expressed as rate of formation of B, is

$$\frac{d(B)}{dt} = k_1 g[(A)_0 - (B)] \tag{2-2}$$

where $(A)_0$ is the amount of A injected, (B) the amount of B produced for contact time, t, k_1 the rate constant for the reaction, and g the fraction of reactant molecules on the surface of the catalyst.

On integration, Eq. (2-2) gives the amount of B as a function of contact time t:

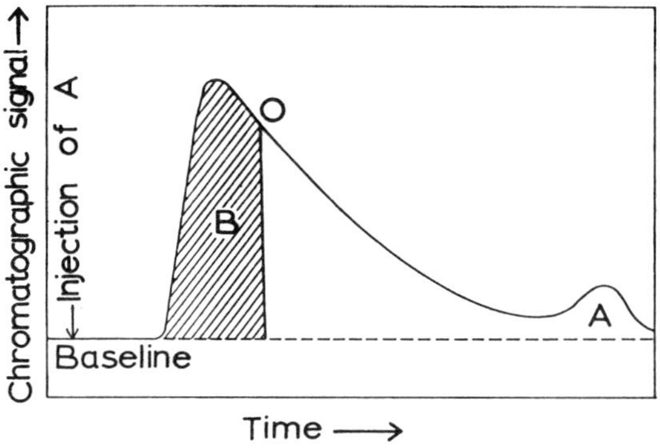

Fig. 2.1 Gas chromatographic trace when a substance A undergoes a first-order reaction giving B, on a column which acts both as a catalytic reactor and as a chromatographic column for separating B from A (2).

$$(B) = (A)_0[1 - \exp(-k_1 gt)] \qquad (2\text{-}3)$$

Eq. (2-3) is easily changed into a linear form to calculate k_1, provided that t can be determined experimentally. The amount of B is relatively easy to measure, since at any point of the chromatographic trace, say O, it is proportional to the area under the curve of B (shown shaded in Fig. 2.1). The amount of A injected is of the order of mm^3 for a liquid, and this cannot be measured accurately. More difficult, however, is the determination of the contact time, t, corresponding to a certain point of the trace, say O, because of the finite retention time of B on the column, and the asymmetry of the peak of B owing to its continuous production from A.

These difficulties can be overcome by <u>stopping</u> the flow of the carrier gas through the column for a definite time interval, Δt. On restoring the flow of the carrier gas, the amount $\Delta(B)$ of B produced during this time interval appears as an extra symmetrical peak superimposed on the tail of the main peak of B, as shown in Fig. 2.2. If Δt is small compared with the half-

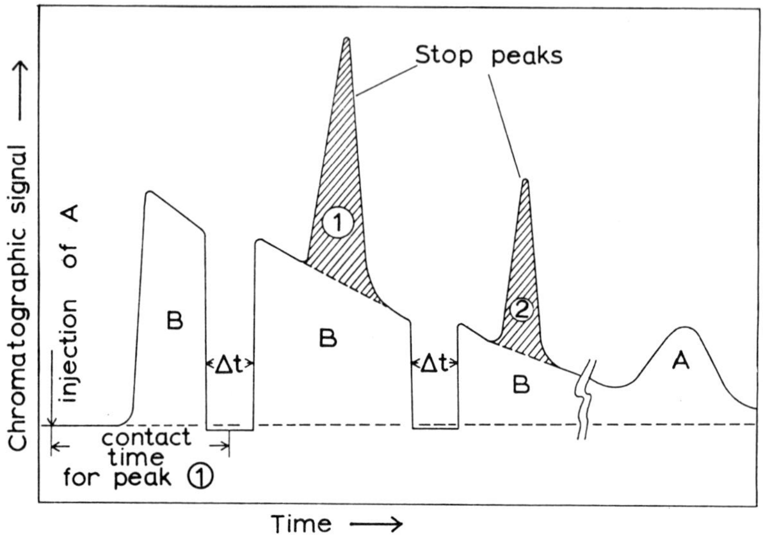

Fig. 2.2 Gas chromatographic trace showing the stopping of the carrier gas through the column for time Δt, when after restoring it the extra stop peaks 1 and 2 appear. The shaded area is proportional to Δ(B) (2).

life of the reaction, the quotient $\Delta(B)/\Delta t$ can be taken, as a first approximation, equal to the rate of formation $d(B)/dt$ of the product B. The relation between the latter and the contact time can be found by differentiating* Eq. (2-3) with respect to t [2]:

$$\frac{\Delta(B)}{\Delta t} \simeq \frac{d(B)}{dt} = (A)_0 k_1 g \exp(-k_1 g t) \qquad (2\text{-}4)$$

Now, Δ(B) is proportional to the area under the curve of the extra "stop peak" produced, Δt is a measurable quantity, and t can be taken equal to the interval Δt. Thus, by repeatedly stopping and restoring the flow of the carrier gas,

*This derivation is different from the original one of Phillips et al. (1).

the rate of the reaction can be determined as a function of time. This contrasts with most methods in chemical kinetics, where a concentration, rather than a rate, is measured as a function of time.

Inspection of Eq. (2-4) shows that a plot of the logarithm of the rate $\Delta(B)/\Delta t$ vs. t should give a straight line from the slope of which k_1g can be found. Having determined g, as described later, the true rate constant, k_1, of the surface reaction can be calculated. For constant stopped-flow intervals Δt, ln [$\Delta(B)$] can be plotted against t, since Eq. (2-4) gives

$$\ln [\Delta(B)] = \ln [(A)_0 k_1 g \Delta t] - k_1 g t \qquad (2\text{-}5)$$

An example is given in Fig. 2.3. Since the stop peaks are usually symmetrical and have the same width at their half-height, the logarithm of their height, instead of their area, can be plotted vs. time.

It is worth noting that the retention time of the stop peaks, measured from the moment of restoring of the carrier gas, does not remain constant, but decreases with increasing number of stops, since the reactant A moves along the column, leaving less and less column length for chromatographing the product B.

The measurement of the retention time of the reactant permits the calculation of g, since it is equal to $[(V_R^o/V_M^o) - 1]/(V_R^o/V_M^o)$, V_R^o being the corrected retention volume of the reactant, and V_M^o the dead volume of the column. The latter can be found by injecting a nonadsorbed gas onto the column, such as air (using a thermal conductivity detector), or methane (when a flame ionization detector is employed).

In some cases, the reacting substance is not eluted from the column, either because its retention volume is too large or becaue the rate of its decomposition is high. In both cases,

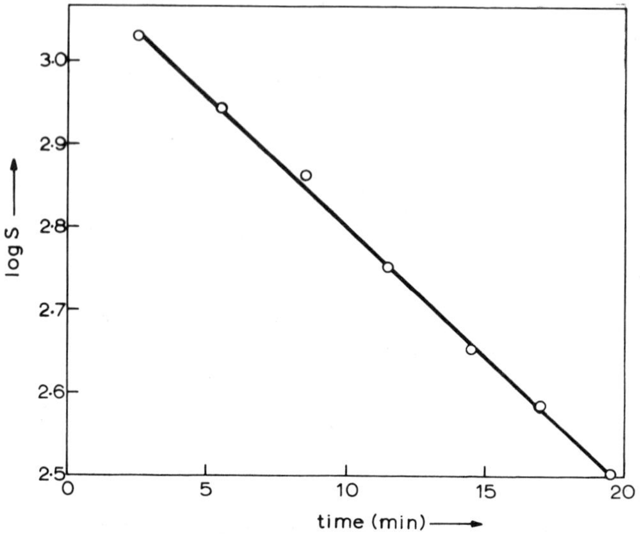

Fig. 2.3 The plot of the logarithm of the area under the stop peaks vs. time, according to Eq. (2-5), for the dehydrohalogenation of 1-bromobutane on 9.1% KBr/Al_2O_3 catalyst at 387.5 K (3).

the retention time of the reactant, necessary to calculate g, can be found indirectly by plotting the retention time of the stop peaks versus the corresponding time from the injection moment to the middle of the relevant stop-flow interval, Δt (after subtracting all previous stop intervals). A straight line results whose x intercept gives the retention time of the reactant A, since this corresponds to zero retention time of the product, i.e., to the time when A would be found at the exit of the column. The correctness of this method has been confirmed in many cases [4].

Determination of true rate constants at various temperatures leads to the calculation of the true activation energy, E_a, and frequency factor, A, for the surface reaction by the use of the well-known Arrhenius equation:

$$\ln k_1 = \ln A - \frac{E_a}{R} \cdot \frac{1}{T} \qquad (2\text{-}6)$$

In Fig. 2.4, some Arrhenius plots are shown from the slopes and intercepts of which E_a and A are easily calculated (3).

II. EXPERIMENTAL SETUP

The stopped-flow method requires only a conventional gas chromatograph, equipped with a suitable detector (usually a flame ionization detector), and modified as shown schematically in Fig. 2.5.

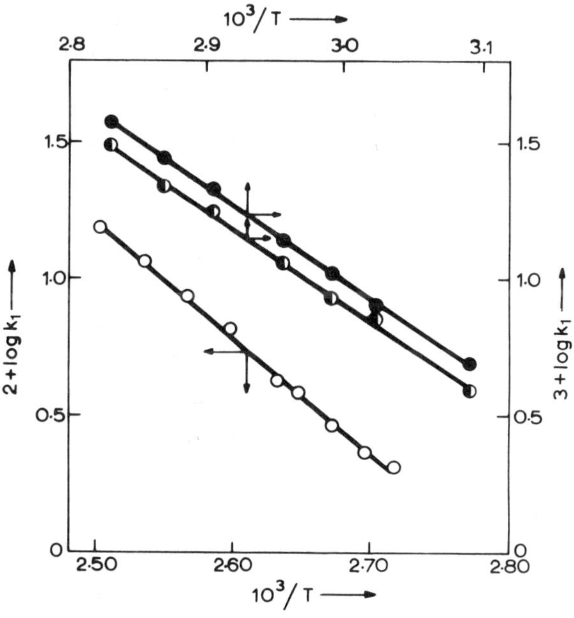

Fig. 2.4 Arrhenius plots for dehydrohalogenation on 9.1% KBr/Al_2O_3 of 1-bromo-2-methylpropane (o), and of 2-bromobutane giving cis-2-butene (•) and trans-2-butene (◐), as studied by the stopped-flow technique (3).

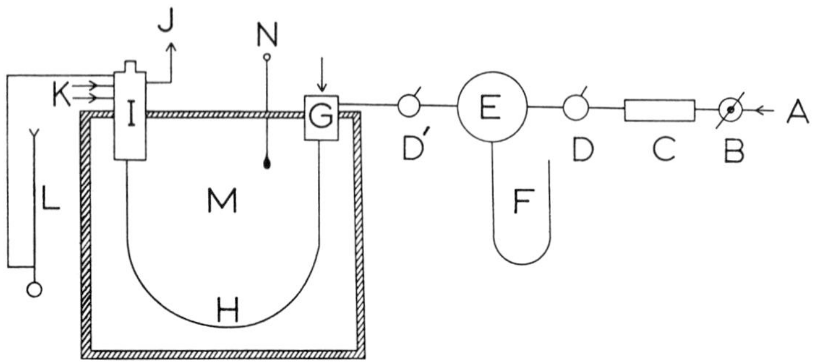

Fig. 2.5 Arrangement for kinetic studies by the stopped-flow technique. A, carrier gas inlet for the catalytic column; B, control valve; C, gas-drying tube; D, D', shut-off valves for closing and opening the carrier gas through the column; E, 500-cm^3 volume reservoir to prevent pressure variations during closing and opening the gas flow; F, open manometer to detect pressure variations and measure the pressure drop along the catalytic column; G, injector to the column heated at oven temperature; H, catalytic column; I, flame ionization detector; K, hydrogen and air to the detector; J, signal to amplifier and recorder; L, bubble flowmeter; M, gas chromatograph oven; N, thermometer (2).

The conditioning of the catalytic column is performed in situ by heating it at an appropriate temperature under continuous carrier gas flow. During this period the end of the column is disconnected from the detector to avoid contamination of the latter. The working temperature is then adjucted and some reactant injections are made to establish constant catalytic activity. The kinetic experiments are performed after about 8h standing at the working temperature.

The end of the column is reconnected to the detector, a small amount of methane is injected to determine the dead volume of the column, following which an injection (0.2 to 1 mm^3) of the liquid-reacting substance is made for kinetic measurements.

These consist in short time (usually 1 min) stops of the carrier gas through the catalytic column by simultaneously closing and opening both valves D and D' (see Fig. 2.5). Each stop is made as soon as the elution of the previous stop peak is complete.

Identification of the product peaks is achieved by injecting small amounts of pure products and comparing their retention times with those of the stop peaks obtained initially.

III. IDEAL STOPPED-FLOW GAS CHROMATOGRAPHY WITH CHEMICAL REACTIONS IN THE STATIONARY PHASE

The derivation of Eq. (2-5), describing the area under the stop peaks as an analytic function of time when the corresponding stop was made (2) as well as an earlier derivation (1), was based on a simple first-order surface reaction (Eq. (2.1)) taking place under static conditions. However, extensive application of the stopped-flow technique has shown that the analytic function mentioned above can be complicated, obviously because of a complex reaction mechanism and/or because of the dynamic character of the chromatographic process in itself. We, therefore, set forth to derive the various chromatographic equations and the above-mentioned function for a fairly general mechanism and a flow system. Then we shall show that the results of the previous derivations (1,2) coincide with the simplest specific case of this mechanism. Other specific cases of the general mechanism are also discussed.

A. General Theoretical Analysis

The following mechanism is assumed to describe most cases of stopped-flow chromatography with chemical reactions taking place on the surface of the column material (5):

$$A + S^{(1)} \rightleftharpoons A - S^{(1)} \underset{k_{-1}^{(1)}}{\overset{k_1^{(2)}}{\rightleftharpoons}} B - S^{(1)} \xrightarrow{k_2^{(1)}} D - S^{(1)} \rightleftharpoons D + S^{(1)}$$

$$A + S^{(2)} \rightleftharpoons A - S^{(2)} \underset{k_{-1}^{(2)}}{\overset{k_2^{(2)}}{\rightleftharpoons}} B - S^{(2)} \xrightarrow{k_2^{(2)}} D - S^{(2)} \rightleftharpoons D + S^{(2)}$$

$$\cdot \quad \cdot \quad \cdot \quad \cdot \quad \cdot \quad \cdot \quad \cdot \quad \cdot \quad \cdot \quad \cdot$$

$$\cdot \quad \cdot \quad \cdot \quad \cdot \quad \cdot \quad \cdot \quad \cdot \quad \cdot \quad \cdot \quad \cdot \qquad (2\text{-}7)$$

$$\cdot \quad \cdot \quad \cdot \quad \cdot \quad \cdot \quad \cdot \quad \cdot \quad \cdot \quad \cdot \quad \cdot$$

$$A + S^{(n)} \rightleftharpoons A - S^{(n)} \underset{k_{-1}^{(n)}}{\overset{k_1^{(n)}}{\rightleftharpoons}} B - S^{(n)} \xrightarrow{k_2^{(n)}} X - S^{(n)} \rightleftharpoons X + S^{(n)}$$

gas surface adsorbed species gases

The surface is assumed to contain one or more kinds of active sites, $S^{(1)}$, $S^{(2)}$, ... , $S^{(n)}$, which are responsible for chromatographing the reactant A and the products D, ... , X, as well as for the chemical reactions of the adsorbed species. The same product, e.g., D, may be produced and/or chromatographed on more than one kind of active sites. The concentration of the various kinds of sites is considered large compared to the concentration of the respective adsorbed species, in view of the very small amounts of the reacting vapors being used.

Additional assumptions are:

1. The adsorption isotherm is linear for all kinds of sites.
2. Axial diffusion of the gases in the bed is negligible, which is not unrealistic for high enough flowrates.
3. The reacting vapor is introduced at the inlet end of the column as an instantaneous pulse, the distribution in time of which can be described by a Dirac delta function $\delta(t)$.

4. Equilibration of the reactants and products between the gas and the solid phases is instantaneous; i.e., the chromatography is ideal.

We artificially divide the time variable in three intervals, t, t_s, and t', as shown in Fig. 2.6. In each interval, the various concentrations as functions of time and distance, x, are determined by one or more differential equations with certain initial and/or boundary conditions. The problem will be considered seperately in each of the three intervals (5).

1. Interval t

The concentrations $c_A(x,t)$, $q_B^{(i)}(x,t)$, and $c_D(x,t)$, where $i = 1, 2, \ldots, n$, are determined by the following system of equations (see List of Symbols).

Mass balance for A and $A - S^{(i)}$:

$$\frac{\partial c_A}{\partial t} + r \sum_{i=1}^{n} \frac{\partial q_A^{(i)}}{\partial t} = -v \frac{\partial c_A}{\partial x} - r \sum_{i=1}^{n} k_1^{(i)} q_A^{(i)} + r \sum_{i=1}^{n} k_{-1}^{(i)} q_B^{(i)} \tag{2-8}$$

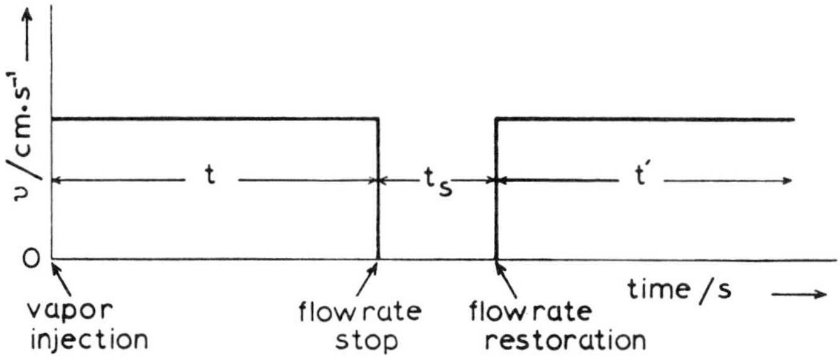

Fig. 2.6 Ideal variation of carrier gas linear velocity, v, with time in the stopped-flow technique (5).

Rate of change of $B - S^{(i)}$:

$$\frac{\partial q_B^{(i)}}{\partial t} = k_1^{(i)} q_A^{(i)} - k_{-1}^{(i)} q_B^{(i)} - k_2^{(i)} q_B^{(i)} \tag{2-9}$$

Mass balance for D and $D - S^{(j)}$:

$$\frac{\partial c_D}{\partial t} + r \sum_j \frac{\partial q_D^{(j)}}{\partial t} = -v \frac{\partial c_D}{\partial x} + r \sum_j k_2^{(j)} q_B^{(j)} \tag{2-10}$$

where index j runs over all sites, producing D and/or equilibrating with it.

Initial and boundary conditions:

$$\text{At } x = 0 \qquad c_A = \frac{m}{\dot{V}} \delta(t) \tag{2-11}$$

At $x > 0$ and $t = 0$, $c_A = q_B^{(i)} = c_D = 0$

Adsorption isotherms:

$$q_A^{(i)} = K_A^{(i)} c_A, \quad q_D^{(j)} = K_D^{(j)} c_D \tag{2-12}$$

By substituting Eqs. (2-12) for $q_A^{(i)}$ and $q_D^{(j)}$ in Eqs. (2-8), (2-9), and (2-10), taking the Laplace transform with respect to t of the resulting three equations under the initial conditions (2-11), and then combining the three transformed equations, we obtain

$$\frac{dC_A}{dx} + \frac{\Omega_A}{v} C_A = 0 \tag{2-13}$$

$$Q_B^{(i)} = \frac{k_1^{(i)} K_A^{(i)}}{p + k_{-1}^{(i)} + k_2^{(i)}} C_A \tag{2-14}$$

$$\frac{dC_D}{dx} + \frac{\Omega_D}{v} C_D = \frac{\Phi_D}{v} C_A \tag{2-15}$$

where

$$\Omega_A = \left(1 + \sum_i k_A^{(i)}\right) p + \sum_i k_1^{(i)} k_A^{(i)} - \sum_i \frac{k_{-1}^{(i)} k_1^{(i)} k_A^{(i)}}{p + k_{-1}^{(i)} + k_2^{(i)}} \tag{2-16}$$

$$\Omega_D = \left(1 + \sum_j k_D^{(j)}\right) p \tag{2-17}$$

$$\Phi_D = \sum_j \frac{k_2^{(j)} k_1^{(j)} k_A^{(j)}}{p + k_{-1}^{(j)} + k_2^{(j)}} \tag{2-18}$$

and $k_A^{(i)} = rK_A^{(i)}$, $k_D^{(j)} = rK_D^{(j)}$ are the partition ratios of A and D, respectively, for the active sites i and j.

The solution of Eq. (2-13) with respct to x, subject to conditions (2-11), is

$$c_A = \frac{m}{\dot{V}} \exp\left(-\frac{\Omega_A x}{v}\right) \tag{2-19}$$

If this is substituted for c_A in Eq. (2-15) and the resulting equation is solved, e.g., by Laplace transformation with respect to x, we obtain

$$c_D = \frac{m}{\dot{V}} \Phi_D \frac{\exp(-\Omega_D x/v) - \exp(-\Omega_A x/v)}{\Omega_A - \Omega_D} \tag{2-20}$$

We are interested in the chromatographic signal of A and D at the detector, which is $c_A(l,t)$ and $c_D(l,t)$ and these can be found by putting $x = l$, so that $x/v = l/v = t_M$ in Eqs. (2-19) and (2-20), and taking the inverse Laplace transforms with respect to p. Thus

$$c_A(l,t) = \mathscr{L}_p^{-1}\left[\frac{m}{\dot{V}} \exp(-\Omega_A t_M)\right] \tag{2-21}$$

$$c_D(l,t) = \mathscr{L}_p^{-1}\left[\frac{m}{\dot{V}} \Phi_D \frac{\exp(-\Omega_D t_M) - \exp(-\Omega_A t_M)}{\Omega_A - \Omega_D}\right] \tag{2-22}$$

These inverse transformations for the general case are a difficult task, but they can be easily found for a number of special or limiting cases, one of which is discussed later. Further, one can find for the general case the characteristics of the elution curve, from the statistical moments, m_n, calculated by means of the well-known property of the Laplace transform:

$$m_n = \mathscr{L}_0^{\infty} t^n c(z,t)dt = (-1)^n \lim_{(p\to 0)} \frac{d^n[C(l,p)]}{dp^n} \tag{2-23}$$

Thus the area under the elution curve of unreacted A is $f_A = \int_0^\infty c_A dV = \dot{V} \int_0^\infty c_A dt \cdot \dot{V} m_0$

$$= m \cdot \exp\left[-\left(\sum_i k_1^{(i)} k_A^{(i)} - \sum_i \frac{k_{-1}^{(i)} k_1^{(i)} k_A^{(i)}}{k_{-1}^{(i)} + k_2^{(i)}}\right) t_M\right] \tag{2-24}$$

Two interesting special cases are when all $k_2^{(i)} = 0$, i.e., no products are formed, and when all $k_{-1}^{(i)} = 0$, i.e., nonopposing reactions occur. In the first case, $f_A = m$; i.e., the whole amount of the injected vapor is eluted. In the second case, $f_A = m \cdot \exp(-\Sigma_i k_1^{(i)} k_A^{(i)} t_M)$; i.e., the amount of the reactant A is diminished according to a first-order law, since $k_A^{(i)} t_M$ is the adjusted retention time $t_{R,A}^{(i)\prime}$, and therefore the contact time the substance would have if only sites i were present.

The mean retention time of A is given by

$$t_{R,A} = \frac{m_1}{m_0} = \left[1 + \sum_i k_A^{(i)} + \sum_i \frac{k_{-1}^{(i)} k_1^{(i)} k_A^{(i)}}{(k_{-1}^{(i)} + k_2^{(i)})^2}\right] t_M \tag{2-25}$$

and therefore depends on the various rate constants. Only when $k_{-1}^{(i)} = 0$ (nonopposing reactions) or $k_1^{(i)} = 0$ (no reaction) is $t_{R,A} = (1 + \Sigma k_A^{(i)})t_M$; i.e., the expected ideal retention time.

2. Interval t_s

During this stopped-flow interval, the various concentrations are again governed by Eqs. (2-8), (2-9), and (2-10) with $v = 0$ and t having been replaced by t_s. The initial conditions are now the inverse Laplace transforms of Eqs. (2-19), (2-14), and (2-20) for $c_A, q_B^{(i)}$ and c_D, respectively.

Although it is not absolutely necessary, we employ an approximation here in order to keep the calculations simple. This approximation, which is justified by the experimental conditions, is to take the interval t_s sufficiently small so that c_A and $q_B^{(i)}$ do not change appreciably during this interval. Then, the only mass balance required here is that of D and D − $S^{(j)}$:

$$\frac{\partial c_D}{\partial t_s} + r \sum_j \frac{\partial q_D^{(j)}}{\partial t_s} = r \sum_j k_2^{(j)} q_B^{(j)} \tag{2-26}$$

After substitution of the isotherm (Eq. (2-12)) for $q_D^{(j)}$ and integration with respect to t_s, we obtain

$$\Delta c_D = c_D - (c_D)_{t_s=0} = \frac{t_s r \sum_j k_s^{(j)} q_B^{(j)}}{1 + \sum_j k_D^{(j)}} \tag{2-27}$$

Thus, Δc_D is the increase in the concentration of D in the gas phase during the stopped-flow interval t_s. It is worth noting that Δc_D is also a function of x and t, having the same distribution in these variables as $q_B^{(j)}$. Therefore, by forcing Δc_D out of the column, one can see the actual distribution of $q_B^{(j)}$ at the time, t, when the stop was made, distorted only by the chromatographic process on D. This is the role of the next time interval.

3. Interval t'

This starts by opening the carrier gas at the end of the stopped-flow interval t_s. Here we are interested only in Δc_D at the detector, i.e., at x = 1. We can imagine that at t' = 0 a distribution in c_D described by Eq. (2-27) is introduced into the column, and it is chromatographed with the same partition ratios $k_D^{(j)}$. Thus we can write the mass balance:

$$\left(1 + \sum_j k_D^{(j)}\right) \frac{\partial(\Delta c_D)}{\partial t'} = -v \frac{\partial(\Delta c_D)}{\partial x} \tag{2-28}$$

A double Laplace transformation first with respect to t' (with initial condition Eq. [2-27]) and then with respect to t gives

$$\frac{d(\overline{\Delta C}_D)}{dx} + \frac{\Omega'_D}{v} \cdot \overline{\Delta C}_D = \frac{t_s r \sum_j k_2^{(j)} Q_B^{(j)}}{v} \tag{2-29}$$

where

$$\Omega'_D = \left(1 + \sum_j k_D^{(j)}\right) p' \tag{2-30}$$

Eq. (2-19) is not substituted for C_A in Eq. (2-14), and the resulting equation is substituted for $Q_B^{(j)}$ in Eq. (2-29). Then, the latter can be integrated with respect to x, giving

$$\overline{\Delta C}_D = \frac{m}{\dot{V}} t_s \Phi_D \cdot \frac{\exp(-\Omega'_D x/v) - \exp(-\Omega_A x/v)}{\Omega_A - \Omega'_D} \tag{2-31}$$

For x = 1, i.e., at the detector, this equation gives the double Laplace transform (with respect to t and t') of the stop peak. It is only necessary to take the inverse transform first with respect to p (included in Φ_D and Ω_A), and then with respect to p' (included in Ω'_D), in order to find $\Delta c_D = \Delta c_D(t, t_s, t')$. This is the extra chromatographic signal above the continuous signal of D, which is due to a stop of the carrier gas flow, made at time t and having duration t_s.

It is worth noting that Eq. (2-31) has exactly the same form as Eq. (2-20), which on inversion gives the continuous chromatographic signal of the product D. The important difference between the two equations, however, is that in Eq. (2-31) the time parameter is split into two independent parameters, p and p', and this makes the inverse Laplace transformation easier. Moreover, one can find the area under the curve of each stop peak (f_s) as a function of the time, t, of the corresponding stop of the carrier gas flow, by using Eq. (2-23):

$$\mathscr{L}_t f_s = \dot{V} \mathscr{L} \int_0^\infty \Delta C_D(p, t')dt' = \dot{V} m_0 = \dot{V} \lim_{(p' \to 0)} (\overline{\Delta C}_D) \tag{2-32}$$

Thus from Eqs. (2-32) and (2-31)

$$f_s = \mathscr{L}_p^{-1} \left[m t_s \Phi_D \frac{1 - \exp(-\Omega_A t_M)}{\Omega_A} \right] \tag{2-33}$$

This is as far as the general case can go. Further development of the equations requires an exact specification of the mechanism as dictated by experimental evidence; e.g., two kinds of active sites and nonopposing reactions forming only one product. Under such specified conditions, the basic equations (2-21) for

c_A, (2-22) for c_D, (2-31) for $\overline{\Delta c_D}$, and (2-33) for f_s, together with the definitions (2-16), (2-17), (2-18), and (2-30) for Ω_A, Ω_D, Φ_D, and Ω'_D, respectively, assume a simpler form and can be manipulated more easily for inverse Laplace transformations with respect to p and p'. Particularly, f_s can be found as an analytic function of t, permitting the calculation of the rate constants of the reactions from experimental data. Some of these specified cases will be discussed below.

B. Nonopposing Reaction on One Kind of Active Site with No Intermediate

This is the simplest possible case of the mechanism 2-7, and we shall show that the ultimate equations coincide with those derived earlier in a static manner (1,2). It is represented by the scheme:

$$A + S \rightleftharpoons A - S \xrightarrow{k_1} D - S \rightleftharpoons D + S \tag{2-34}$$

and is expressed by $k_A^{(2)}, k_A^{(3)}, \ldots, k_A^{(n)} = 0$, $k_D^{(2)}, k_D^{(3)}, \ldots, k_D^{(n)} = 0$, $k_{-1}^{(i)} = 0$ (i = 1, 2, . . . , n) and $k_2^{(i)} = \infty$ (i = 1, 2, . . . , n).

Under these condiions, $\Omega_A = (1 + k_A)p + k_1k_A$, $\Omega_D = (1 + k_D)p$, $\Phi_D = k_1k_A$, and $\Omega'_D = (1 + k_D)p'$. Eq. (2-21) then becomes

$$c_A(l,t) = \mathscr{L}_p^{-1}\left\{\frac{m}{\dot{V}} \exp(-k_1k_At_M) \cdot \exp[-(1 + k_A)t_Mp]\right\}$$

$$= \frac{m}{\dot{V}} \exp(-k_1k_At_M) \cdot \delta[t - (1 + k_A)t_M] \tag{2-35}$$

or since from Eq. (2-25) $t_{R,A} = (1 + k_A)t_M$, and $k_At_M = t'_{R,A}$ (adjusted retention time),

$$c_A(l,t) = \frac{m}{\dot{V}} \exp(-k_1t'_{R,A}) \cdot \delta(t - t_{R,A}) \tag{2-36}$$

This is the expected equation for linear ideal chromatography with the signal at the detector being diminished because of the factor $\exp(-k_1 t'_{R,A})$, describing a first-order reaction with contact time $t'_{R,A}$.

Eq. (2-22) becomes

$$c_D(1,t) = \mathscr{L}_p^{-1} \left\{ \frac{m}{\dot{V}} k_1 k_A \cdot \frac{\exp[-(1+k_D)t_M p] - \exp(-k_1 k_A t_M) \cdot \exp[-(1+k_A)t_M p]}{(k_A - k_D)p + k_1 k_A} \right\}$$

$$= \frac{m}{\dot{V}} k_1 \lambda \{\exp[-k_1 \lambda (t - t_{R,D})] \cdot u(t - t_{R,D})$$

$$- \exp(-k_1 t'_{R,A}) \cdot \exp[-k_1 \lambda (t - t_{R,A})] \cdot u(t - t_{R,A})\} \tag{2-37}$$

where $t_{R,D} = (1 + k_D)t_M$ is the ideal retention time the product D would have, if it were injected directly onto the column,

$$\lambda = \frac{k_A}{k_A - k_D} = \frac{t'_{R,A}}{t'_{R,A} - t'_{R,D}} = \frac{t'_{R,A}}{t_{R,A} - t_{R,D}} \tag{2-38}$$

and $u(t - t_R)$ is the Heaviside unit step function, which is 0 for $t < t_R$ and 1 for $t \geq t_R$, with $t_R > 0$.

For $t < t_{R,D} < t_{R,A}$ both terms within the braces { } of Eq. (2-37) are zero and $c_D = 0$, i.e., no signal is recorded at the detector. The first signal should appear abruptly at $t = t_{R,D} < t_{R,A}$, when the first term becomes 1, whereas the second remains 0. The maximum in the signal is therefore $(c_D)_{max} = mk_1 \lambda/\dot{V}$; i.e., inversely proportional to $\dot{V}$. For $t_{R,D} < t < t_{R,A}$ the signal decreases exponentially with t, so that from Eq. (2-37)

$$\ln c_D = \ln\left(\frac{mk_1 \lambda}{\dot{V}}\right) + k_1 \lambda t_{R,D} - k_1 \lambda t \tag{2-39}$$

Thus a plot of the logarithm of the height of the elution curve of D versus t should be linear with slope $-k_1 \lambda$. This conclusion coincides with the simple elution technique of Phillips et al. [1].

Finally, when $t > t_{R,A}$ both terms in Eq. (2-37) come into play, and again $c_D = 0$, as can be seen by substituting for λ the expression on the far right of Eq. (2-38) and performing the calculations.

We consider next the most important Eq. (2-31), which reduces to (for x = 1):

$$\overline{\Delta C}_D = \frac{m}{\dot{V}} t_s k_1 k_A \cdot \frac{\exp(-t_{R,D}p') - \exp(-k_1 t'_{R,A}) \cdot \exp(-t_{R,A}p)}{(1 + k_A)p + k_1 k_A - (1 + k_D)p'} \qquad (2\text{-}40)$$

Taking the inverse Laplace transform first with respect to p and then with respect to p', we find for $t < t_{R,A}$:

$$\Delta c_D(t, t_s, t') = \frac{m}{\dot{V}} t_s k_1 g \exp(-k_1 g t) \cdot \delta\left[t' - t_{R,D}\left(1 - \frac{t}{t_{R,A}}\right)\right] \qquad (2\text{-}41)$$

where g is the fraction of the reactant A in the adsorbed form A − S:

$$g = \frac{k_A}{1 + k_A} \qquad (2\text{-}42)$$

and $(1 + k_D)/(1 + k_A)$ has been replaced by $t_{R,D}/t_{R,A}$.

Eq. (2-41) predicts the elution of a stop peak, since Δc_D has a non-zero value only for

$$t' = t_{R,D}\left(1 - \frac{t}{t_{R,A}}\right) \tag{2-43}$$

which thus is the retention time of the stop peak. This retention time does not remain constant but decreases as t increases and becomes zero when $t = t_{R,A}$. Only when $t \ll t_{R,A}$ is $t' \simeq t_{R,D}$, i.e., the retention of the stop peak is approximately equal to the ideal retention time of the product D only when the stop is made early in the run or when the retention time of the reactant is very large.

Finally, we find the area under the curve of the stop peak as a function of t using Eq. (2-33), which here becomes

$$f_s = \mathscr{L}_p^{-1}\left[mt_s k_1 k_A \frac{1 - \exp(-k_1 t'_{R,A}) \cdot \exp(-t_{R,A}p)}{(1 + k_A)p + k_1 k_A}\right]$$

giving

$$f_s = mt_s k_1 g \cdot \exp(-k_1 g\, t)[1 - u(t - t_{R,A})] \tag{2-44}$$

where the expression in brackets [ ] is 1 for $t < t_{R,A}$ and 0 for $t > t_{R,A}$. This result coincides with Eq. (2-5) previously found using another derivation (2).

A plot of the logarithm of the stop-peak area versus t is predicted linear with slope $-k_1 g$. This has been confirmed experimentally in certain cases (1–3,6–8).

C. Nonopposing Reactions on Two Kinds of Active Sites

If in the general mechanism described by Eq. (2-7) only two kinds of sites, 1 and 2, are assumed to be active, all partition ratios of the reactant $k_A^{(i)}$, and of the product $k_D^{(j)}$, with i,j = 3, 4, . . . , n are equal to zero. Furthermore, if nonopposing reactions are considered, all $k_{-1}^{(i)} = 0$ (i = 1, 2, . . . , n). Under these conditions, the functions Ω_A, Φ_D, and Ω'_D, defined by Eqs. (2-16), (2-18), and (2-30), respectively, take the form:

$$\Omega_A = (1 + k_A^{(1)} + k_A^{(2)})p + k_1^{(1)}k_A^{(1)} + k_1^{(2)}k_A^{(2)} \qquad (2\text{-}45)$$

$$\Phi_D = \frac{k_2^{(1)}k_1^{(1)}k_A^{(1)}}{p' + k_2^{(1)}} + \frac{k_2^{(2)}k_1^{(2)}k_A^{(2)}}{p + k_2^{(2)}} \qquad (2\text{-}46)$$

$$\Omega'_D = (1 + k_D^{(1)} + k_D^{(2)})p' \qquad (2\text{-}47)$$

We are interested in Eqs. (2-31) and (2-33), the first of which gives (for x = 1, i.e., at the detector) the double Laplace transform of the stop peak(s) with respect to t and t'. By working out the inverse Laplace transform of this equation, one can find the elution curve of the stop peak(s) Δc_D as a function of t and t'. It turns out that Δc_D has non-zero values only in the narrow interval

$$t_{R,D}\left(1 - \frac{t}{t_{R,A}}\right) < t' < t_{R,D} \qquad (2\text{-}48)$$

where $t_{R,A}$ and $t_{R,D}$ (the conventional retention times for the reactant and the product, respectively) are given by the relations:

$$t_{R,A} = \frac{(1 + k_A^{(1)} + k_A^{(2)})l}{v}$$

$$t_{R,D} = \frac{(1 + k_D^{(1)} + k_D^{(2)})l}{v} \qquad (2\text{-}49)$$

Thus the elution of a narrow stop peak with a retention time approximately equal to that of the product is predicted. How narrow this peak is depends on the time when the carrier-gas flow was stopped and the retention time of the reactant $t_{R,A}$. If the latter is large, the stop peaks are very narrow and their retention time does not change appreciably with t.

The second equation of interest is Eq. (2-33), which gives the area under the curve of each stop peak. For the present special case of mechanism (2-7), this equation takes the form:

$$f_s = \mathscr{L}_p^{-1}\left\{mt_s\left[\frac{k_2^{(1)}k_1^{(1)}g^{(1)}}{p + k_2^{(1)}} + \frac{k_2^{(2)}k_1^{(2)}g^{(2)}}{p + k_2^{(2)}}\right]\right.$$

$$\left.\cdot\frac{1 - \exp\left[-\left(k_1^{(1)}k_A^{(1)} + k_1^{(2)}k_A^{(2)}\right)t_M\right]\cdot\exp\left(-t_{R,A}p\right)}{p + \bar{k}_1}\right\} \qquad (2\text{-}50)$$

where $g^{(1)}$ and $g^{(2)}$ are the fractions of reactant A adsorbed on sites 1 and 2, respectively, given by the relations

$$g^{(1)} = \frac{k_A^{(1)}}{1 + k_A^{(1)} + k_A^{(2)}} \qquad g^{(2)} = \frac{k_A^{(2)}}{1 + k_A^{(1)} + k_A^{(2)}} \qquad (2\text{-}51)$$

and

$$\bar{k}_1 = k_1^{(1)}g^{(1)} + k_1^{(2)}g^{(2)} \qquad (2\text{-}52)$$

i.e., an adsorption weighted mean of k_1. On performing the inverse Laplace transformation with respect to p, one finds f_s as an explicit function of t. Ignoring terms which vanish for $t < t_{R,A}$ because they contain as a factor the Heaviside unit step function $u(t - t_{R,A})$, and setting the rate of formation R of the product proportional to f_s/t_s, with proportionality constant α, we finally obtain

$$R = \alpha \frac{f_s}{t_s} = \alpha m \left\{ \frac{k_2^{(1)} k_1^{(1)} g^{(1)}}{\bar{k}_1 - k_2^{(1)}} \exp(-k_2^{(1)} t) + \frac{k_2^{(2)} k_1^{(2)} g^{(2)}}{\bar{k}_1 - k_2^{(2)}} \exp(-k_2^{(2)} t) - \left[\frac{k_2^{(1)} k_1^{(1)} g^{(1)}}{\bar{k}_1 - k_2^{(1)}} + \frac{k_2^{(2)} k_1^{(2)} g^{(2)}}{\bar{k}_1 - k_s^{(2)}} \right] \exp(-\bar{k}_1 t) \right\} \tag{2-53}$$

This equation, derived theoretically on the basis of the mechanism:

$$A + S^{(1)} \rightleftharpoons A - S^{(1)} \xrightarrow{k_1^{(1)}} B - S^{(1)} \xrightarrow{k_2^{(1)}} D - S^{(1)} \rightleftharpoons D + S^{(1)}$$

$$A + S^{(2)} \rightleftharpoons \underbrace{A - S^{(2)} \xrightarrow{k_1^{(2)}} B - S^{(2)} \xrightarrow{k_2^{(2)}} D - S^{(2)}} \rightleftharpoons D + S^{(2)} \tag{2-54}$$

gas surface adsorbed species gas

describes very well the rate of catalytic deaminations on aluminum oxide, and specifically the time dependence of the experimental rate of formation of cyclohexene from aminocyclohexane (9) and dicyclohexylamine (10) as well as the rate of formation of butenes from acyclic amines (9). Empirically it has been observed that this reaction obeys an equation of the form:

$$R = a_1 \exp(-b_1 t) + a_2 \exp(-b_2 t) - a_3 \exp(-b_3 t) \tag{2-55}$$

and it is easy to show that Eq. (2-53) has exactly this form in the following cases: 1. when $\bar{k}_1 > k_2^{(1)}$ and $\bar{k}_1 > k_2^{(2)}$; 2. when $k_2^{(1)} > \bar{k}_1 > k_2^{(2)}$ and the first preexponential factor on the r.h.s. of Eq. (2-53) is absolutely greater than the second; 3. when $k_2^{(2)} > \bar{k}_1 > k_2^{(1)}$ and the second preexponential factor is absolutely greater than the first. In all three cases, the term with the greatest exponential coefficient is negative and the value of its preexponential factor should be equal to the sum of the values of the other two preexponential factors. This has been confirmed experimentally, within the limits of experimental error (9).

It is admitted that Eq. (2-55) and the equivalent theroetical Eq. (2-53) have six adjustable aprameters and one might argue that with this number of parameters a fit of the experimental results would easily be obtained. The practice, however, shows that this is not the case because of the required restrictions and relations between these parameters predicted theoretically above, and supported by experiment. Also, the fact that among other models tried only that of Eq. (2-54) explains satisfactorily all experimental findings, which further supports the proposed mechanism on two sites. The existence of the latter on the alumina surface has been suggested by many authors in connection with various types of reactions. The possibility of a self-poisoning effect of the reactant has been excluded by the fact that the results were, within the limits of experimental error, independent of the number of amine injections. Anyhow, the amount of the reactant in the stopped-flow technique is so small compared with the amount of the catalyst that even if a self-poisoning effect exists, it would not influence the kinetic data obtained.

Figure 2-7 gives an example of behavior described by Eq. (2-55) or (2-53). It also shows how the preexponential factors a_1, a_2, a_3 and the exponential coefficients b_1, b_2, b_3 can be

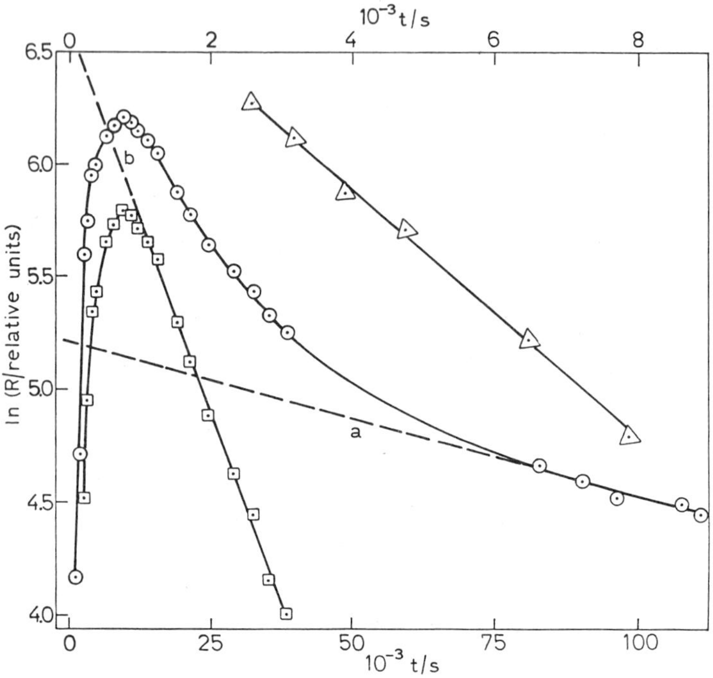

Fig. 2.7 Time dependence of the experimental rate of formation (R) of cyclohexene by deamination of aminocyclohexane on aluminum oxide of pH 7, at 495 K, under a volume flow rate ($\dot{V}$) of carrier gas 0.173 $cm^3 \cdot s^{-1}$.

The rate R was taken proportionally to the height of the stop peaks in cm as mentioned in the procedure, whereas t is the time interval from the injection of amine to the beginning of the stopped-flow interval: o (lower abscissa), experimental points R; □ (lower abscissa), values of R' obtained by subtracting from the experimental points R the corresponding values $a_1 \exp(-b_1 t)$ given by the linear extrapolation (a); Δ (upper abscissa), values of R" obtained by subtracting R' from the corresponding values $a_2 \exp(-b_2 t)$ given by the linear extrapolation (b) (9).

determined from the experimental curve. The latter coefficients coincide with the rate constants $\bar{k}_1$, $k_2^{(1)}$ and $k_2^{(2)}$ of Eq. (2-53).

From the variation of these constants with temperature activation energies and frequency factors were determined using conventional Arrhenius plots (9,10). These are true and not apparent activation energies, since $\bar{k}_1$, $k_2^{(1)}$, and $k_2^{(2)}$ refer to the decomposition of adsorbed species, not involving heats of adsorption.

D. Nonopposing Reactions on Three Kinds of Active Sites

The experimental results in the catalytic deamination of acyclic amines on porous glass (9), and of aminocyclohexane on silica gel (11), and also in the dehydration of cyclopentanol and cyclohexanol on irradiated aluminum oxide (12) can be explained by assuming that a third kind of active site, $S^{(3)}$, is operative in addition to $S^{(1)}$ and $S^{(2)}$. Then, a third series of reactions must be added to Eq. (2-54):

$$A + S^{(3)} \rightleftharpoons A - S^{(3)} \xrightarrow{k_1^{(3)}} B - S^{(3)} \xrightarrow{k_2^{(3)}}$$

$$D - S^{(3)} \rightleftharpoons D + S^{(3)} \tag{2-56}$$

and all relevant equations modified accordingly. We limit ourselves in the final Eq. (2-53), which here becomes

$$R = \alpha m \left\{ \sum_{i=1}^{3} \frac{k_2^{(i)} k_1^{(i)} g^{(i)}}{\bar{k}_1 - k_2^{(i)}} \exp\left(-k_2^{(i)} t\right) - \left[\sum_{i=1}^{3} \frac{k_2^{(i)} k_1^{(i)} g^{(i)}}{\bar{k}_1 - k_2^{(i)}} \right] \exp\left(-\bar{k}_1 t\right) \right\} \tag{2-57}$$

where $g^{(i)}$ and $\bar{k}_1$ are now given by

$$g^{(i)} = \frac{k_A^{(i)}}{1 + \sum_{i=1}^{3} k_A^{(i)}} \qquad \bar{k}_1 = \sum_{i=1}^{3} k_1^{(i)} g^{(i)} \tag{2-58}$$

If the third stage of Eq. (2-56), $B - S^{(3)} \rightarrow D - S^{(3)}$, is very fast so that $k_2^{(3)} \rightarrow \infty$, the term with i = 3 in the first summation of Eq. (2-57) becomes zero, whereas the term with i = 3 in the second summation becomes $-k_1^{(3)}g^{(3)}$. Thus this equation reduces again to Eq. (2-53), with the only difference that a term $-k_1^{(3)}g^{(3)}$ is added within the brackets [ ] of its r.h.s., making the preexponential factor of the last term smaller (absolutely) than the sum of the preexponential factors of the other two terms, and possibly smaller than either of them. This is in accordance with experimental findings in deaminations on porous glass [9]. It is thus noteworthy that although a third type of active sites is postulated, the new equation has again the <u>same</u> number of adjustable parameters, though differing in their physical meaning.

E. Catalytic Conversions Calculated by Stopped-Flow Gas Chromatography

The deamination reactions on aluminum oxide and on porous glass (9) mentioned in the previous two paragraphs require that the first summation in Eq. (2-57) has only two terms. This can happen in three ways: 1. All $k_2^{(i)}$ with $i \geq 3$ tend to ∞ or, in physical terms, the rate determining step on these sites is the formation of the intermediate $B - S^{(i)}$. 2. All $k_1^{(i)}g^{(i)}$ for $i \geq 3$ are zero (i.e., the amine adsorbs and reacts only on sites 1 and 2); on all other kinds of sites either it adsorbs reversibly ($g^{(i)} \neq 0$), but it does not react ($k_1^{(i)} = 0$), or it does not adsorb at all ($g^{(i)} = 0$). 3. For $i \geq 3$ $k_1^{(i)}g^{(i)} \neq 0$, but

$k_2^{(i)} = 0$ (i.e., the amine adsorbs and transforms irreversibly to B − $S^{(i)}$ but this does not yield gaseous products).

In all studies published so far in which the stopped-flow technique was employed, the various rate constants, k, were accurately determined from the slopes of plots of ln R against t; however, for the preexponential factors of Eq. (2-57)

$$a^{(i)} = \frac{k_2^{(i)} k_1^{(i)} g^{(i)}}{\bar{k}_1 - k_2^{(i)}} \qquad (2\text{-}59)$$

only <u>relative</u> values were determined. These relative values permit us to decide whether possibility (1) above is substantiated or not, but they do not allow us to distinguish between possibilities (2) and (3). This is because in both last cases, the preexponential factor of the negative term in Eq. (2-57) is equal to the sum of the values of the other two preexponential factors, and this has been found to hold for deaminations on aluminum oxide (9). In contrast, in case (1) above, the second summation of Eq. (2-57) contains terms $-k_1^{(i)} g^{(i)}$ with $i \geq 3$, which makes the summation smaller (absolutely) than the sum of $a^{(i)}$'s of the other two terms. This happens in deaminations on porous glass (9).

The distinction between possibilities (2) and (3) is of some importance, since it affects not the rate constants of the reactions, but the extent of <u>conversion</u> of amine into the desired product. This can be seen by integrating with respect to t the expression in braces { } on the r.h.s. of Eq. (2-57), between the limits 0 and ∞, obtaining unity in case (2); i.e., 100% conversion, whereas in case (3) the result is:

$$\text{conversion} = \frac{k_1^{(1)} g^{(1)} + k_1^{(2)} g^{(2)}}{\sum_{i=1}^{n} k_1^{(i)} g^{(i)}} < 1 \qquad (2\text{-}60)$$

i.e., less than 100%. Experimentally, this can be done either calculating the conversion directly from the chromatogram, or by finding the absolute values of the preexponential factors $a^{(i)}$, calculating from them $k_1{}^{(i)}g^{(i)}$ by Eq. (2-59) and then computing the conversion by means of Eq. (2-60). The denominator of (2-60) is $\bar{k}_1$, according to Eq. (2-58). The latter method, i.e., through the absolute $a^{(i)}$ values, will demonstrate the internal consistency of the theory underlying Eq. (2-57), and also its potentialities for calculating not only rate constants, but also conversions through the rate constants.

These further potentialities of the stopped-flow technique have been demonstrated (13) in a specific case; i.e., the deamination of aminocyclohexane on aluminum oxide, where the experimental data bear out possibilities (2) or (3) mentioned above. The absolute values of the preexponential factors $a^{(i)}$ given by Eq. (2-59) have been determined using the relation:

$$a_{abs}^{(i)}/S^{-1} = (a_{rel}^{(i)}/cm)\,\frac{(w_{1/2}/s)1.064}{(m/mol)(t_s/s)(S/cm\ s\ mol^{-1})} \tag{2-61}$$

$a_{rel}{}^{(i)}$ are the relative values of the preexponential factors of Eq. (2-57), as determined from the intercepts of ln(h/cm) against time plots, $w_{1/2}$ is the half-width of the stop peaks, 1.064 is a factor for finding the area under the stop peaks from their height, h, and their half-width, and S the response of the detector.

The effective value of t_s does not necessarily coincide with the real time during which the valves of the carrier gas are closed, as measured by a watch, and it can be calculated as follows. After injecting 0.5 mm^3 of amine and performing several stops of the carrier gas until the chromatographic signal decays to a negligible height, the total amount of products is given by the integral

$$A = \int_0^\infty R dt = \int_0^\infty \frac{1.064\ hw_{1/2}}{t_s} dt = \frac{1.064 w_{1/2}}{t_s} \int_0^\infty h dt \tag{2-62}$$

where R is the rate of the reaction, and t the time of stopping the carrier gas. The value of A is found from the total area under the elution curve, which can be measured by a disc integrator. The integral on the far right of Eq. (2-62) can be found by numerical integration with respect to time of the stop-peak height, h, above the continuous elution curve. Since $w_{1/2}$ is known from the chromatogram, t_s can be calculated.

Finally, the response S in Eq. (2-61) can be found by injecting known amounts of the pure product into an empty column and integrating the resulting elution curve.

The next two steps are to calculate $k_1^{(i)} g^{(i)}$ from $a_{abs}^{(i)}$ using Eq. (2-59), and then the conversion by means of Eq. (2-60).

From the results found (13), the conclusion can be drawn that the rate equation (2-57) is internally consistent in that it does not only explain the time dependence of the rate for the deamination reactions, permitting the calculation of rate constants, but it also through its preexponential factors correlates these rate constants with the total conversion of the reactant to products. This is accomplished by finding the absolute values of the intercepts of Eq. (2-57) by means of Eq. (2-61). The calculated conversions are in good agreement with those found experimentally.

From the conversions found or calculated, details of the mechanism emerge, i.e., that deamination takes place on two kinds of active sites of aluminum oxide, but a considerable fraction of the reacting amine adsorbs on other kinds of sites and transforms irreversibly to other adsorbed products (case (3) mentioned before).

All previous paragraphs B to E are based on specific cases of the general mechanism (2-7). Another branched mechanism was postulated to explain the kinetics of n-butene isomerization on molecular sieve 13X, silica gel, and alumina, as studied by the stopped-flow technique (14). Based on this mechanism, equations analogous to Eqs. (2-31) and (2-33) were derived following a similar procedure, and then used to analyze the experimental data. The interested reader may refer to the original paper (14) for the details of the derivation and calculations.

IV. NONIDEAL STOPPED-FLOW GAS CHROMATOGRAPHY

One of the assumptions of the previous section III.A was that equilibration of the solutes between the gas and the solid phases is instantaneous; i.e., that chromatography is ideal. In the present section, we shall discuss briefly the stopped-flow technique in nonideal, i.e., nonequilibrium, gas chromatography, but without any chemical reaction in the stationary phase. Only adsorption-desorption phenomena between the gas and the stationary phase will be assumed. It had been shown earlier (15) that the stopped-flow technique can be used to measure desorption rates in nonequilibrium gas chromatography. This was based on a simple theoretical analysis with a homogeneous surface, but the experimental data for desorption of various hydrocarbons from an alumina surface suggested the existence of two kinds of adsorption sites with two different desorption rates. The idea of a two-sited adsorption-desorption was further explored (16), and it has been shown that not only two desorption rate constants but also the two corresponding partition coefficients can be extracted from the experimental data.

It has been observed that the stopped-flow technique can with advantage be applied in adsorption-desorption studies if the adsorbate is first allowed to equilibrate between the gas and the solid phases at the inlet end of the chromatographic column, although this is not always necessary. Thus equilibrium initial conditions are used here, which differ from those of conventional pulse chromatography.

Consider a conventional tubular gas chromatographic column filled with a solid adsorbent. Under isothermal conditions, a small amount of the pure adsorbate A is instantaneously introduced as vapor at the inlet end of the column and allowed to equilibrate with the solid. Carrier gas flow through the column (at a constant velocity) is now established, destroying the equilibrium between the gas and the solid phases, and giving a highly asymmetric signal at the detector. This signal has a sharp front profile and a long tailing.

While the chromatographic signal is decaying, the flow of the carrier gas is stopped and restored after a definite time interval, and this is repeated, noting the time of each stop. Following each restoring of the carrier gas flow, a sharp symmetric peak (the so-called stop peak) is detected, having a well-defined retention time and superimposed on the otherwise asymmetric tailing signal, as shown in Fig. 2.8. The stop peak is due to the enrichment in adsorbate of the gas phase owing to slow desorption during the nonflow interval. The problem to be solved again is to find the area under the curve of each stop peak as a function of the time of the corresponding stop of the carrier gas flow, under the following assumptions:

1. Axial diffusion in the bed is negligible.
2. The only slow processes determining rates of equilibration of the adsorbate between the gas and the solid phases are adsorption and desorption phenomena, other resistances, e.g., intraparticle diffusion, being negligible.

The above two assumptions seem reasonable for high enough flowrates and small particle diameters.

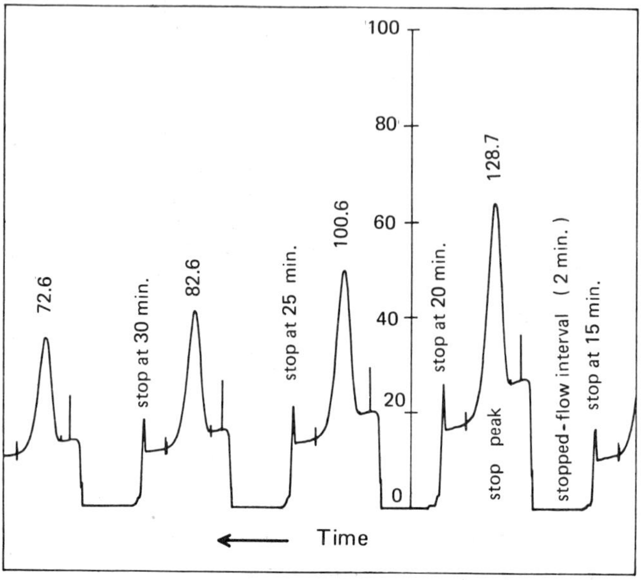

Fig. 2.8 Section of a typical stopped-flow chromatogram for desorption of n-heptane from an alumina surface covered with 10% potassium chloride at 111°C. Nitrogen ws used as carrier gas with a corrected volume flowrate of 0.43 cm^3/s. The number on each stop peak is the area of the peak in mV · s. (16).

3. Adsorption-desorption is taking place on two kinds of active sites, $S^{(1)}$ and $S^{(2)}$. This is dictated by experimental evidence. The adsorption isotherm is assumed linear for both kinds of sites, their fractions (s_1 and s_2) of the total concentration of sites remaining constant with time.

This is not an unreasonable assumption for small adsorbate concentrations.

4. The adsorbate is introduced in an infinitesimally small section of the column, so that the feed band can be described by a Dirac delta function $\delta(x)$. The use of a square function, such as the difference of two Heaviside step functions $u(x) - u(x - b)$, is more complicated, but leads to equivalent results.

For the adsorption-desorption of the adsorbate A on the two kinds of reactive sites we can write:

$$A + S^{(1)} \underset{k_{-1}}{\overset{k_1}{\rightleftharpoons}} A - S^{(1)} \tag{2-63}$$

$$A + S^{(2)} \underset{k_{-2}}{\overset{k_2}{\rightleftharpoons}} A - S^{(2)} \tag{2-64}$$

where $A - S^{(1)}$ and $A - S^{(2)}$ are the two adsorbed species.

Provided that the above are elementary reactions, the equilibrium constants for adsorption are related to the rate constants through the equations:

$$K_1 = \frac{k_1}{k_{-1}} s^{(1)} \tag{2-65}$$

$$K_2 = \frac{k_2}{k_{-2}} s^{(2)} \tag{2-66}$$

As in section III.A, the problem can be considered separately for the three intervals t, t_s, and t' of the time variable. By writing the mass balance in the gas phase and the rates of adsorption on sites $S^{(1)}$ and $S^{(2)}$, a system of partial differential equations results, which can be solved by using Laplace transformations under the given initial conditions. This is done in all time intervals (t, t_s, and t'). Without going into the details of the mathematical treatment, which can be found in the original paper [16], we quote only the final result, obtained by using certain approximations, and giving the area f_s under the curve of the stop peaks:

$$f_s = Y_1 \exp(-k_{-1}t) + Y_2 \exp(-k_{-2}t) \tag{2-67}$$

The functions Y_1 and Y_2 are such that

$$Y_1 + Y_2 = \frac{mrt_s(k_{-1}K_1 + k_{-2}K_2)}{1 + r(K_1 + K_2)} \cdot \exp\left[-(k_{-1}K_1 + k_{-2}K_2)\frac{V_s}{\dot{V}}\right] \quad (2\text{-}68)$$

and

$$\frac{Y_1}{Y_2} = \frac{k_{-1}K_1}{k_{-2}K_2} \cdot \frac{1 - (k_{-2}K_2V_s/\dot{V}) \cdot (k_{-1} + k_{-2})/(k_{-1} - k_{-2})}{1 + (k_{-1}K_1V_s/\dot{V}) \cdot (k_{-1} + k_{-2})/(k_{-1} - k_{-2})} \quad (2\text{-}69)$$

The area of each peak recorded by the instrument is αf_s, where f_s is given by Eq. (2-67) and α is a proportionality constant. The actual value of this constant is not required for the calculation of k_{-1}, k_{-2}, K_1, and K_2 with the help of Eqs. (2-67), (2-68), and (2-69), as has been described [16]. Alternatively, one can normalize each stop-peak area by dividing it by the total area under the elution curve of the adsorbate, as recorded by a second integrator. The result is a relative (dimensionless) area, f_{rel}, which is equal to f_s/f_{total}.

According to Eq. (2-67), a plot of ln f_{rel} against t will yield a curve, and if k_{-1} and k_{-2} are sufficiently different, the plot will become linear after a certain time, because the term with the greatest value of k (conventionally taken as k_{-1}) becomes negligibly small. An example is given in Fig. 2.9. The slope of the last linear section gives $-k_{-2}$ and its intercept with the y axis equals ln (Y_2/f_{total}). A second straight line is constructed by plotting against t the logarithm of the points calculated from the difference between the experimental points and those found by extrapolation of the first straight line. The slope of the second line gives $-k_{-1}$ and its y intercept equals

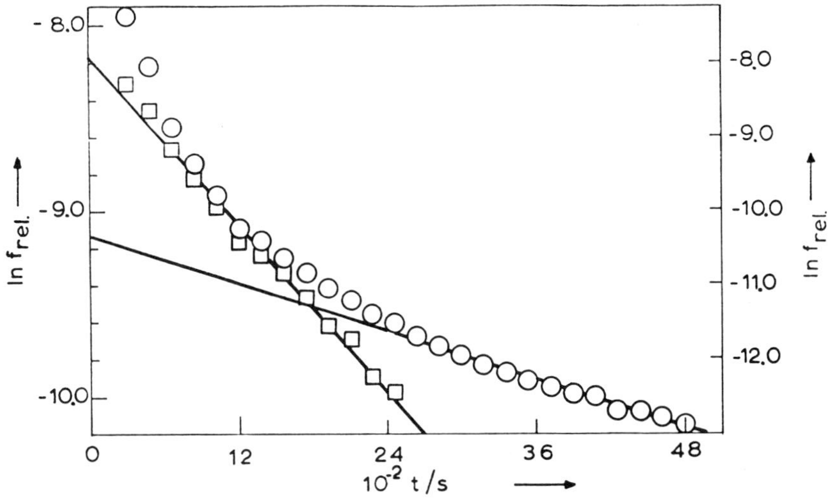

Fig. 2.9 Experimental data for desorption of heptane from modified alumina at 396.3 K and with 0.315 $cm^3 \cdot s^{-1}$, plotted according to Eq. (2-67); o, experimental points (left ordinate); □, points calculated from the difference between the experimental points and those found by extrapolation of the last linear section (right ordinate) (17).

ln (Y_1/f_{total}). The two rate constants of desorption k_{-1} and k_{-2} are thus determined.

At a particular temperature, the two equilibrium constants K_1 and K_2, can be determined from kinetic runs with different flowrates, $\dot{V}$, by plotting ln $[(Y_1 + Y_2)/f_{total}]$ against $1/\dot{V}$. According to Eq. (2-68) this plot should again be linear with slope $-(k_{-1}K_1 + k_{-2}K_2)V_s$, where V_s is known and k_{-1}, k_{-2} have already been determined. We thus have an algebraic equation in K_1 and K_2. A second equation is provided by Eq. (2-69), where the ratio Y_1/Y_2 is equal to the ratio $(Y_1/f_{total})/(Y_2/f_{total})$ determined from experimental values. These two equations are then solved to give the values of K_1 and K_2.

The above procedure was used for the determination of rate constants for desorption, and of equilibrium constants for ad-

Table 2.1 Activation Energies and Activation Entropies for Desorption of Heptane from Both Kinds of Active Sites, 1 and 2, of Al_2O_3 - 10% w/w KCl and of Porous Glass

	E_a/kJ mol^{-1}		$\Delta S^{\ddagger}$/J K^{-1} mol^{-1}	
Surface	Site 1	Site 2	Site 1	Site 2
Al_2O_3-10% KCl	15.1 ± 0.8	18.8 ± 0.4	−269 ± 3	−277 ± 1
Porous glass	13.8 ± 0.8	10.5 ± 0.4	−270 ± 3	−307 ± 1

Source: Ref. (17).

sorption of heptane on alumina modified with 10% KCl and on porous glass (17). Experiments at various temperatures have permitted the calculation of desorption activation energies and entropies of activation from the variation of the rate constants with temperature as well as the calculation of adsorption enthalpies and entropies from the variation of equilibrium constants with temperature. The results are shown in Tables 2.1 and 2.2.

V. MEASUREMENT OF DIFFUSION COEFFICIENTS IN GASES

Gas chromatographic methods for measuring diffusion coefficients of a gas A into another gas B are known for over 2 decades, and a review has been published (18) by Maynard and

Table 2.2 Enthalpies and Entropies of Adsorption of Heptane on Both Kinds of Active Sites, 1 and 2, of Al_2O_3-10% w/w KCl and of Porous Glass

	ΔH/kJ mol^{-1}		ΔS/J K^{-1} mol^{-1}	
Surface	Site 1	Site 2	Site 1	Site 2
Al_2O_3-10% KCl	− 7.2 ± 0.3	−21 ± 4	43.1 ± 0.8	8 ± 2
Porous glass	−35 ± 2	−29 ± 4	−42 ± 4	−17 ± 3

Source: Ref. (17).

Grushka. It covers the theme of zone broadening by diffusion of a narrow pulse of component A (the solute) introduced into a long empty chromatographic column, through which component B is continuously flowing as a conventional carrier gas. A variation of the method [19] uses the extra broadening caused by stopping the carrier gas flow for varying time periods. Stopped-flow gas chromatography also has been employed more recently [20], not for a mere broadening of an existing chromatographic zone, but to create new very narrow peaks (stop peaks) on an asymmetrical elution curve of A. The decay of these stop peaks with time was used to determine the diffusion coefficient of A into B, with much shorter columns. The decisive column part can be made straight, so that secondary flow phenomena due to coiled tubes are avoided.

Consider a conventional tubular empty GC column, at one end of which a small volume of the pure component A is instantaneously introduced, as a gas or vapor, in the form of a pulse (by means of a syringe or a gas valve). The other component, B, fills the whole column but flows continuously only through a part of it, by entering, not at the point of injection of A, but at an intermediate position down the column, as shown in Fig. 2.10. This arrangement creates a discontinuity in the concentration gradient of A at the inlet point of B, i.e., at $z = L$ or $x = 0$. This is because equality of fluxes of A through a cross section of the column at $z = L$ requires that the flux in the z side due to diffusion be balanced by the flux in the x side due mainly to the bulk velocity of B. Once this discontinuity is established, a <u>short stop</u> in the flow of B will produce an accumulation of A at $x = 0$, which after restoration of the B flow will be recorded by the detector as an extra stop peak, as shown in Fig. 2.11. This can be repeated giving a series of stop peaks, the height (and the area) of which diminishes with time. The problem to be solved is to determine

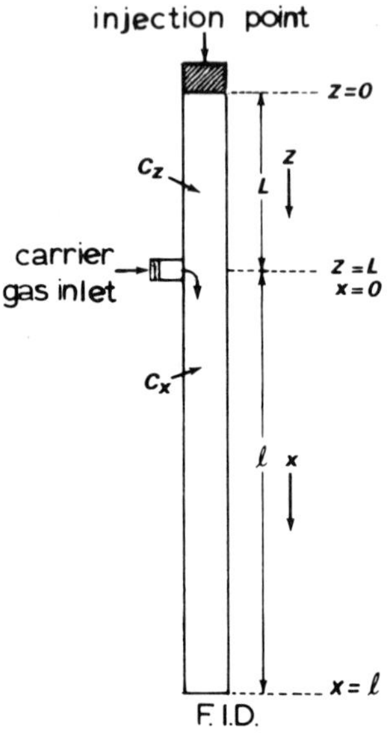

Fig. 2.10 Schematic representation of the diffusion (L) and the chromatographic (l) columns for measuring diffusion coefficients (20).

the area under the curve of each stop peak as a function of the time of the corresponding stop in the flow of the carrier gas B.

The following assumptions are made:

1. Radial diffusion in the column is negligible.
2. Axial diffusion of A in the region x of the column is negligible. This seems reasonable for a high enough flowrate of B.
3. The solute A is introduced in an infinitesimally small section of the column region, z, so that the feed band can be described by a delta function $\delta(z)$.

The time variable can again be divided in three intervals, t, t_s, and t', and the problem considered separately in each of these intervals. Without going into the mathematical details

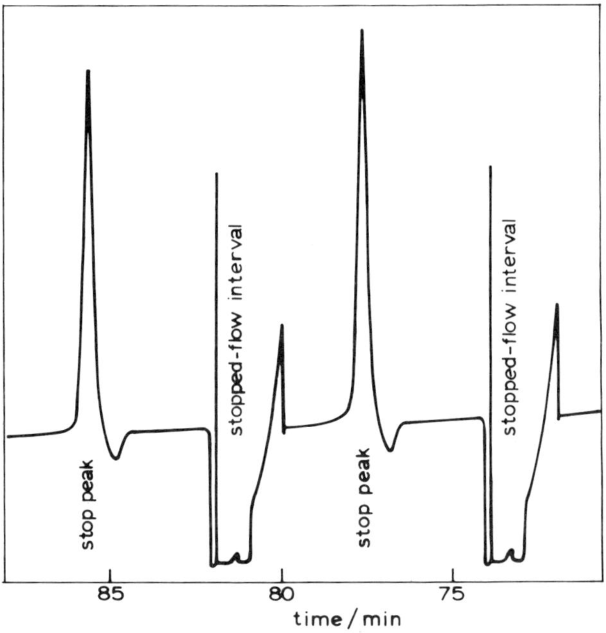

Fig. 2.11 A stopped-flow chromatogram for measuring diffusion coefficients. The solute was propene and the carrier gas nitrogen. (L = 40 cm, l = 2.7 m, $\dot{V}$ = 0.167 $cm^3 s^{-1} N_2$, T = 296 K, p = 1 atm) (20).

which have been published [20], we give only the final equation for the area under the stop peaks as a function of time:

$$f_s = \frac{m t_s L}{(\pi D)^{1/2} t^{3/2}} \cdot \exp\left(-\frac{L^2}{4Dt}\right) \tag{2-70}$$

From this, a linear form is obtained:

$$\ln (f_s t^{3/2}) = \ln \left(\frac{m t_s L}{\pi^{1/2} D^{1/2}}\right) - \frac{L^2}{4D} \cdot \frac{1}{t} \tag{2-71}$$

which permits the calculation of D from the slope ($-L^2/4D$) of the ln ($f_s t^{3/2}$) plot versus 1/t. Since the stop peaks are fairly

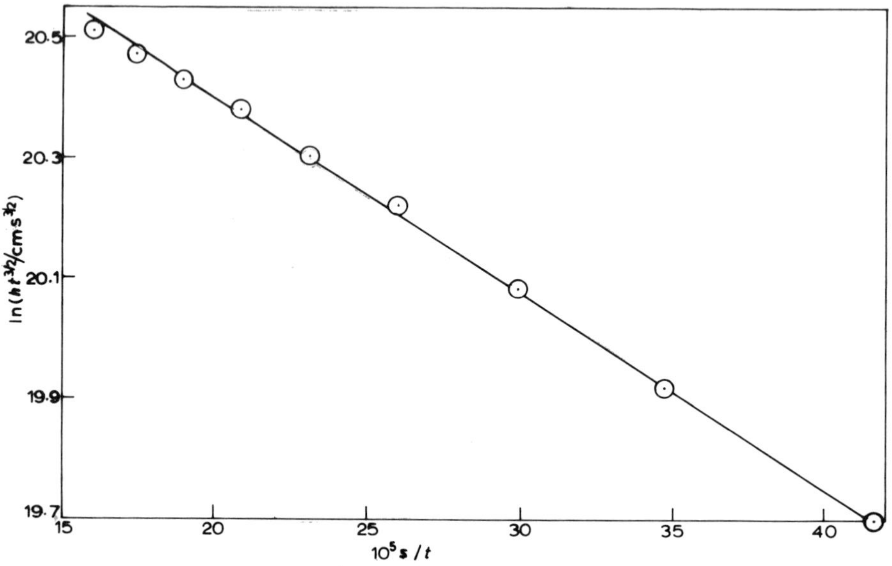

Fig. 2.12 Plot of diffusion data of propene into nitrogen obtained by the stopped-flow method, according to Eq. (2-71). The height, h, of the stop peaks from the baseline is used, as mentioned in the text (L = 40 cm, 1 = 2.7 m, $\dot{V}$ = 0.167 $cm^3s^{-1}N_2$, T = 296 K, p = 1 atm) (20).

symmetrical and have a constant half-width, their height from the baseline rather than their area, f_s, can be used to plot Eq. (2-71). This avoids complicated integrations of the stop peaks on a baseline continuously changing in slope.

A typical plot is shown in Fig. 2.12.

LIST OF SYMBOLS

$a^{(i)}$	Preexponential factors defined by Eq. (2-59)
c_A, c_D	Concentrations of A and D in the gas phase
C_A, C_D	Laplace transforms of c_A and c_D with respect to t
D	Mutual diffusion coefficient of two gases

Δc_D	Increase in the gas phase concentration of D during the stopped-flow interval
ΔC_D	Laplace transform of Δc_D with respect to t
$\overline{\Delta C_D}$	Double Laplace transform of Δc_D with respect to t and t'
f_A	Area under the elution curve of A
f_s	Area under the curve of a stop peak
g	Fraction of A in the adsorbed form
$g^{(1)}, g^{(2)}, g^{(3)}$	Fractions of A adsorbed on sites 1, 2, and 3, respectively
h	Height of a stop peak
k_1, k_2	Rate constants for adsorption on sites 1 and 2, respectively
k_{-1}, k_{-2}	Rate constants for desorption from sites 1 and 2, respectively
$k_1^{(i)}, k_{-1}^{(i)}, k_2^{(i)}$	Rate constants for reaction on site i
$\bar{k}_1$	Mean rate constant defined by Eq. (2-52) or (2-58)
$k_A^{(i)}, k_D^{(j)}$	Partition ratios of A and D on site i or j
l	Length of column
L	Diffusion column (cf. Fig. 2.10)
m	Total amount of injected A
m_n	Integrals defined by Eq. (2-23)
p, p'	Transform parameters with respect to t and t'
$q_A^{(i)}, q_B^{(i)}, q_D^{(j)}$	Concentrations of adsorbed species $A - S^{(i)}$, $B - S^{(i)}$, $D - S^{(j)}$ on site i or j per unit volume of solid
$Q_B^{(i)}$	Laplace transform of $q_B^{(i)}$ with respect to t
$s^{(1)}, s^{(2)}$	Fractions of adsorption sites 1 and 2, respectively
S	Response of the detector
r	Volume ratio of solid and gas phases

R	Rate of formation of product
t	Time interval from the injection of solute to the beginning of the stopped-flow interval
t_M	Gas hold-up time
$t_{R,A}, t_{R,D}$	Retention times of A and D
t_s	Stopped-flow interval
t'	Time measured from the end of the stopped-flow interval
v	Linear velocity of the carrier gas in the interparticle space
V	Volume of the carrier gas passed through the column
V_s	Total volume of solid adsorbent
$\dot{V}$	Volume flowrate of carrier gas
V_R^o	Corrected retention volume
V_M^o	Dead volume of the column
$w_{1/2}$	Half-width of a stop peak
x	Distance from inlet end of column
Y_1, Y_2	Functions defined by Eqs. (2-68) and (2-69)
α	Proportionality constant
λ	Expression defined by Eq. (2-38)
Φ_D	Expression defined by Eq. (2-18)
Ω_A, Ω_D, Ω'_D	Expressions defined by Eq. (2-16), (2-17), and (2-30), respectively

REFERENCES

1. C. S. G. Phillips, A. J. Hart-Davis, R. G. L. Saul, and J. Wormald, J. Gas Chromatogr., 5:424 (1967).
2. N. A. Katsanos and A. Lycourghiotis, Chim. chronika, New series, 5:137 (1976).
3. I. Hadzistelios, H. J. Sideri-Katsanou, and N. A. Katsanos, J. Catal., 27:16 (1972).

4. A. Lycourghiotis, Thesis, University of Patras, Patras, 1974.

5. N. A. Katsanos, J. Chromatogr., 152:301 (1978).

5. A. Lycourghiotis, N. A. Katsanos, and I. Hadzistelios, J. Catal., 36:385 (1975).

7. A. Lycourghiotis and N. A. Katsanos, React. Kin. Catal. Lett., 4:221 (1976).

8. R. M. Lane, C. B. Lane, and C. S. G. Phillips, J. Catal., 18:281 (1970).

9. A. Lycourghiotis, N. A. Katsanos, and D. Vattis, J. Chem. Soc., Faraday Trans. I, 75:2481 (1979).

10. G. Karaiskakis, A Lycourghiotis, D. Vattis, and N. A. Katsanos, React. Kin. Catal. Lett., 5:453 (1976).

11. A. Lycourghiotis, D. Vattis, and N. A. Katsanos, Z. Physik. Chem. (N.F.), 126:259 (1976).

12. A. Tseremegli, N. A. Katsanos, and I. Hadzistelios, Z. Physik. Chem. (N.F.), 129:259 (1982).

13. D. Vattis, N. A. Katsanos, G. Karaiskakis, A. Lycourghiotis, and M. Kotinopoulos, J. Chromatogr., 214:171 (1981).

14. N. A. Katsanos and A. Tsiatsios, J. Chromatogr., 213:15 (1981).

15. N. A. Katsanos and I. Hadzistelios, J. Chromatogr., 105:13 (1975).

16. N. A. Katsanos, G. Karaiskakis, and I. Z. Karabasis, J. Chromatogr., 130:3 (1977).

17. G. Karaiskakis and N. A. Katsanos, J. Chromatogr., 151:291 (1978).

18. V. R. Maynard and E. Grushka, Adv. Chromatogr., 12:99 (1975).

19. J. H. Knox and L. McLaren, Anal. Chem, 36:1477 (1964).

20. N. A. Katsanos, G. Karaiskakis, D. Vattis, and A. Lycourghiotis, Chromatographia, 14:695 (1981).

3

The Reversed-Flow Technique

I. Introduction

II. General Experimental Setup

III. Reversed-Flow GC as a Sampling Technique

IV. The Chromatographic Sampling Equation

A. Initial F-interval

B. R-interval

C. Second F-interval

V. Calculation of Rate Coefficients and Equilibrium Constants from the Sample Peaks

List of Symbols

References

I. INTRODUCTION

There are some severe limitations of the stopped-flow technique of gas chromatography (GC) described in the previous chapter. These are: 1. When used for catalytic studies, the reactant must have a relatively long retention time on the column containing the catalyst. 2. It does not seem to apply to reactions other than those with one reactant and first-order steps. For example, it is difficult to apply the stopped-flow method in hydrogenation reactions with complicated kinetics, like methanation of carbon monoxide. From the point of view of physical principles, there is also a small drawback of this technique: It continuously switches the system under study from a flow dynamic one to a static system and vice versa, by repeatedly closing and opening the carrier gas flow. Longitudinal diffusion and other related phenomena, which are usually negligible during the gas flow, may become important when the flow is stopped.

In this chapter, we introduce the second important flow perturbation method; i.e., the reversed-flow technique (1). A preliminary form of this (2) was used to measure the rate constants for the cracking of cumene, the dehydration of propan-1-ol, and the deamination of 1-aminopropane over a 13X molecular sieve. This was followed by a detailed theoretical analysis and application of the method to the dehydration of alcohols, the deamination of primary amines, and the cracking of cumene (3–7).

The method was then extended to more complicated reactions with two reactants (8—12), and to other physicochemical measurements. The latter include adsorption equilibrium constants (13), gas diffusion coefficients (14—17), the kinetics of drying of catalysts (18), rate coefficients for evaporation of pure liquids (19), molecular diameters and critical volumes of gases (20), activity coefficients in liquid mixtures (21), mass transfer coefficients of hydrocarbons on solid adsorbents (22), and for evaporation of liquids (23), Lennard-Jones parameters (24), and finally, interaction between the components of salt-modified adsorbents (25).

II. GENERAL EXPERIMENTAL SETUP

As explained in Chapter 1, section VI, the reversed-flow technique is based on reversing the direction of flow of the carrier gas from time-to-time. The experimental setup for the application of the method is very simple and comprises generally:

1. A conventional gas chromatograph with any kind of detector capable of detecting the vapor(s) contained in the carrier gas. A high-sensitivity device, like a flame ionization detector (FID), or a flame photometric detector (FPD) is to be preferred whenever appropriate. The oven of the chromatograph must be big enough to accommodate two columns if possible, although for most applications, a single column oven will do.

2. A so-called <u>sampling column</u> constructed from chromatographic tube (glass, stainless steel, or other material) of any diameter (usually 1/4 in.), and having a total length 0.8-2.5 m, depending on the particular application. This sampling column can be completely empty of any solid material (in applications like those described in Chapters 4 and 5), or it can be filled with a usual chromatographic material, a catalyst, or it can contain both (cf. Chapter 6). The sampling column is coiled and accommodated inside the chromatograph's oven.

3. A <u>diffusion column</u> constructed from the same material as the sampling column and connected perpendicularly to it, usually at its middle point. The other end of the diffusion column is closed with an injector septum and is sometimes bent to form a U-shaped tube. The whole column can be empty of any solid material (in applications like those described in Chapter 4), or it can be filled with a solid adsorbent (cf. Chapter 5). The diffusion column is relatively short (30-100 cm) and can be straight, protruding out of the chromatograph's oven, or coiled and placed inside the oven with the closed end at the injector position of the chromatograph. When out of the oven, the column is kept at room temperature, or it can be heated by a heating band or by water circulation around it from a thermostat. Examples of particular arrangements applied to

various types of measurements will be found in later chapters of the book.

4. The sampling and the diffusion column described in paragraphs 2 and 3 above form what we call the sampling cell, and this cell must now be connected to the carrier gas inlet and the detector in such a way that the carrier gas flow through the sampling column (no carrier gas flows through the diffusion column) can be reversed in direction at any time desired. This can be done by using a four-port valve to connect the ends D_1 and D_2 of the sampling column to the carrier gas supply and the detector, as shown schematically in Fig. 3.1. With the valve in the position indicated by the solid lines, the carrier gas (dried and regulated by a gas-flow controller) enters the column at D_2 and leaves it from D_1 toward the detector. By switching the valve to the other position (dashed lines), the direction of the carrier gas flow is reversed, entering now the column at D_1.

To avoid condensation of solutes inside the valve and in the 1/16-in. connection tubes, it is preferable to install the valve in such a way that its body is placed inside a large hole of the oven wall and the connection tubes run as far as possible inside the oven. Also, the whole valve body can be placed in a heated enclosure like those being sold in the market.

Instead of a four-port valve, one can use a usual six-port gas sampling valve, with a short 1/16-in. tube connecting two alternate ports, as shown in Fig. 3.2.

5. Whenever a detector with a flame is used (e.g., FID), a restrictor is placed before the detector to prevent the flame from being extinguished when the valve is turned from one position to the other. A restrictor is also needed when one wishes to increase the pressure inside the sampling cell.

Separation of various components contained in the carrier gas is usually effected by filling the sampling column with an

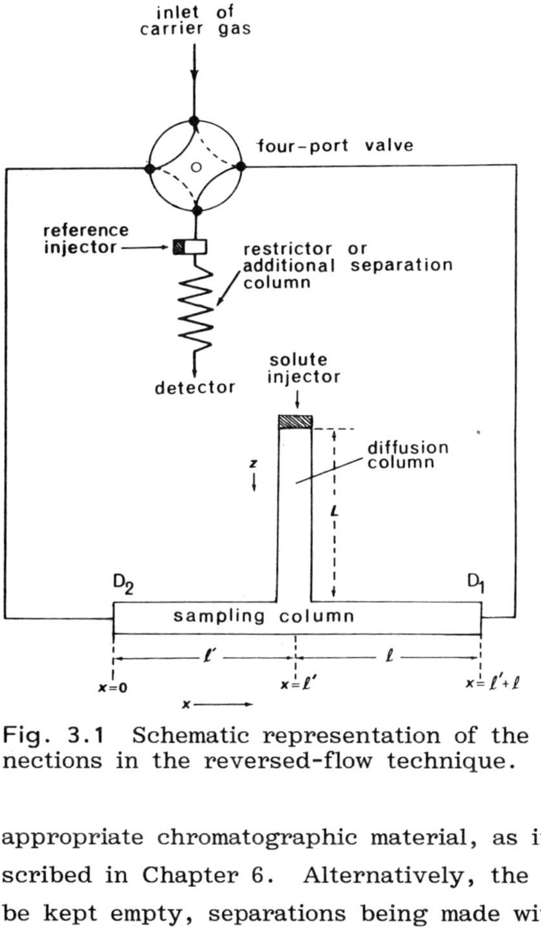

Fig. 3.1 Schematic representation of the columns and gas connections in the reversed-flow technique.

appropriate chromatographic material, as in the applications described in Chapter 6. Alternatively, the sampling column can be kept empty, separations being made with an additional chromatographic column connected in place of the restrictor, which is not needed now (cf. Fig. 3.1). This additional column can be accommodated in the same oven as the sampling column or in a separate oven and heated at a different temperature.

It is to be noted that the flow reversals are confined in the sampling column, whereas in the separation column, the carrier gas flows always in the same direction. Reference substances for identification purposes can be introduced in the

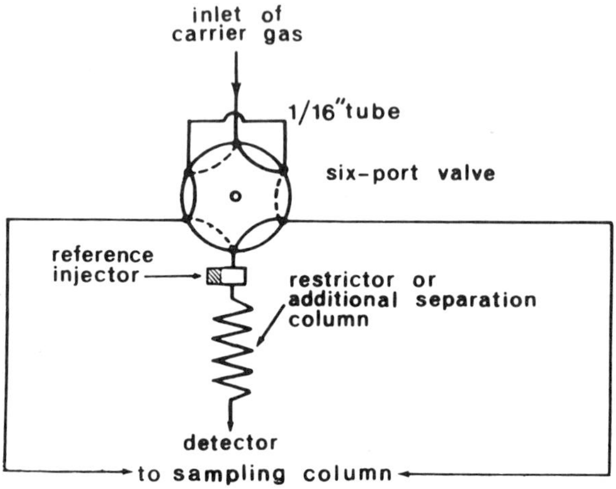

Fig. 3.2 Schematic representation of the columns and gas connections using a six-port valve.

separation column through the reference injector shown in Fig. 3.1.

III. REVERSED-FLOW GC AS A SAMPLING TECHNIQUE

The question now naturally arising is: What we would observe on the chromatographic elution curve when we reverse the direction of the carrier gas flowing in a system like that depicted in Fig. 3.1? If pure carrier gas were passing through the sampling column, nothing would happen on reversing the flow. But if the carrier gas contains other gases at finite concentrations recorded by the detector system, the flow reversals create perturbations on the elution curve, having the form of extra chromatographic peaks. We call these sample peaks, and give an example in Fig. 3.3. A simple flow reversal from one direction to the other creates sample peaks like that marked A, whereas two successive flow reversals, i.e., reversal of the

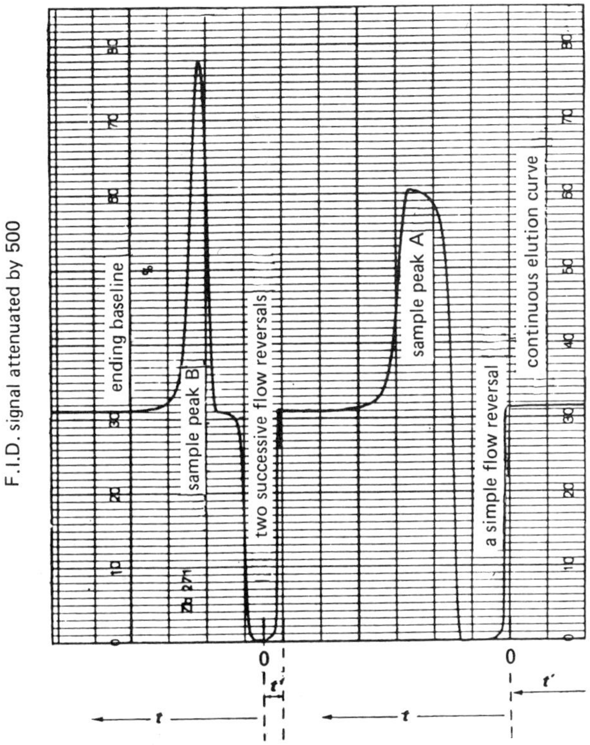

Fig. 3.3 Sample peaks created by reversing the flow direction of carrier gas helium containing small amounts of n-heptane.

flow for a small time period, t', and then restoration of it in its original direction gives rise to peaks like B. These double reversal sample peaks are usually more symmetrical, bigger in height, and can be made as narrow as we want. It will be shown in the next section of this chapter that theoretically the width of the B peaks at their half-height is equal to the duration, t', of the backward flow of the carrier gas.

The loading of the carrier gas with other substances can be due to: 1. processes occurring outside the sampling cell of Fig. 3.1, e.g., by passing the gas through a saturator. In cases like this, the sample peaks created by the flow reversals are nothing else but pulses of the substance contained in the carrier gas, like those which would be obtained if we injected small volumes of this substance through the reference injector; 2. processes taking place within the cell, e.g., the slow diffusion of a gas A contained in the diffusion column into the carrier gas B passing through the sampling column. The enrichment of B in the gas A depends on the rate with which A enters the sampling column at the junction x = l' of the two columns. By reversing now the flow we perform a sampling of the concentration of A at this junction, each sample peak measuring (by its height or its area under the curve) this concentration at the time of the flow reversal. Repeating this sampling procedure at various times and using suitable mathematical analysis, the rate coefficient of the slow process responsible for the sample peaks can be determined; e.g., the diffusion coefficient of A into B in the example given above.

The continuous monitoring of the concentration of other substances in the carrier gas may reveal an equilibrium state in which these substances are involved. In certain cases, the mathematical analysis may permit the calculation of the relevant equilibrium constant.

The rate process or the equilibrium state mentioned above can occur either inside the diffusion column (empty or filled), or inside the sampling column filled with solid material(s). Sometimes these processes are confined to a very short section near the junction of the two columns, but in other cases can take place along the whole length of the diffusion column.

The mathematical analysis permitting the calculation of rate coefficients and equilibrium constants is based on a general equation, describing the elution curve of the sample peaks (1,

15). The derivation of this equation is now given analytically in the next section.

IV. THE CHROMATOGRAPHIC SAMPLING EQUATION

The following derivation is made by reference to Fig. 3.4, which depicts only the sampling column of the cell of Fig. 3.1, supposedly filled with a chromatographic material. The assumptions under which the sampling equation is derived are:

1. The adsorption isotherms in the sampling column are linear.
2. Axial diffusion of the gases along coordinate x or x' and other phenomena leading to nonideality (such as mass transfer resistances in the stationary phase, nonequivalent flow paths in the packed bed) are negligible. This seems reasonable for high enough flow rates.
3. The rate process is taking place in a sufficiently short section of the total column length, so that its x distribution can be described approximately by a Dirac delta function, $\delta(x - l')$.

The problem will be considered separately for various time intervals, in which the concentration of a representative gase-

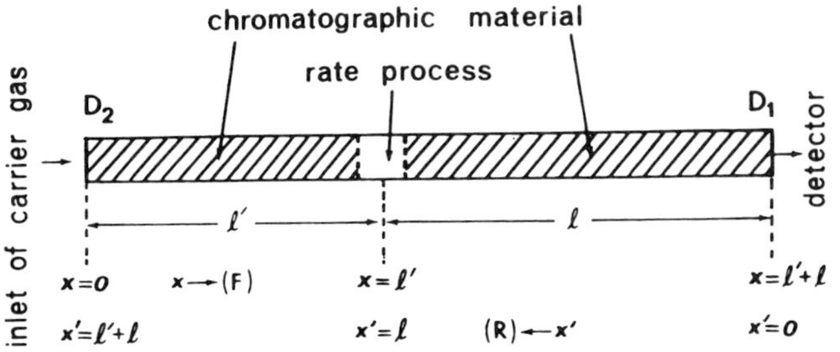

Fig. 3.4 The sampling column of Fig. 3.1, filled with a chromatographic material.

ous component A (as a function of time and distance, x or x') is determined by certain differential equations with given initial and boundary conditions.

The definition of the various symbols used is given in the List of Symbols appended at the end of the chapter.

A. Initial F-Interval

When the carrier gas B flows in the direction F, i.e., from D_2 toward D_1 (see Fig. 3.4), the concentration of the component A at x = l', $c(l',t_0)$, owing to the rate process taking place there, spreads out in the column section, l, thus becoming a function of time and distance, $c(x,t_0)$. This is described by the following mass balance equation, under assumptions (1) and (2):

$$(1 + k_A)\frac{\partial c}{\partial t_0} = - v \frac{\partial c}{\partial x} + v\ c(l',t_0)\ \delta(x - l') \qquad (3\text{-}1)$$

Taking the t_0 Laplace transform of this equation, under the initial condition $c(x,0) = 0$, we find an ordinary differential equation for C as a function of x. This can be integrated by means of its x Laplace transform, with the result:

$$C = C(l',p_0)\ \exp\,(- p_0\,\theta) \cdot (x - l') \qquad (3\text{-}2)$$

where

$$\theta = \frac{(1 + k_A)(x - l')}{v} \qquad (3\text{-}3)$$

and $u(x - l')$ is the Heaviside unit step function, which equals 0 for $x < l'$ and 1 for $x \geq l'$. At the detector, i.e., at $x = l' + l$, $u(x - l')$ becomes $u(l) = 1$ for $l > 0$, and θ becomes $(1 + k_A)l/\upsilon = t_R$, i.e., the ideal retention time of component A in the column section l. Thus Eq. (3-2) gives C at the detector as

$$C = C(l',p_0)\ \exp\,(- p_0 t_R) \qquad (3\text{-}4)$$

and, according to the property translation of Laplace transformations, the inverse transform of Eq. (3-4) for $t_R \geq 0$ is

$$c = c(l', t_0 - t_R) \cdot u(t_0 - t_R) \tag{3-5}$$

This equation describes the break-through curve of component A at position D_1 (Fig. 3.4).

B. R-Interval

While the carrier gas is flowing in direction F, giving the curve described by Eq. (3-5), its flow is reversed to the direction R at a time $t_0 > t_R$. The time measured from the moment of reversal is called t'. The distance coordinate, x, is now changed to x', according to the obvious relation

$$x' = l' + l - x \tag{3-6}$$

and the concentration of A in this time interval, c', becomes a function of x' and t', $c' = c'(x', t')$. It is given by an equation analogous to Eq. (3-1)

$$(1 + k_A) \frac{\partial c'}{\partial t'} = - v \frac{\partial c'}{\partial x'} + v c'(l, t_0, t') \, \delta(x' - l) \tag{3-7}$$

$c'(l, t_0, t')$ now being the concentration due to the rate process at $x' = l$, which is a function of both t_0 and t'.

As with Eq. (3-1), we proceed by taking Laplace transforms of this equation with respect to time, but now the t_0 transform is taken first, and then the t' transform with initial condition

$$C'(x', p_0, 0) = C'(l, p_0, 0) \cdot \exp(p_0 \theta')[1 - u(x' - l)] \tag{3-8}$$

This is obtained from Eq. (3-2) by substituting $1 - u(x' - l)$ for $u(x - l')$ and replacing $-\theta$, as defined by Eq. (3-3), by its equivalent:

$$\theta' = \frac{(1 + k_A)(x' - 1)}{v} = -\theta \qquad (3\text{-}9)$$

The result of the above double Laplace transformation is

$$\frac{d\bar{C}'}{dx'} + \frac{(1 + k_A)p'}{v}\bar{C}' = \frac{1 + k_A}{v} \cdot C'(1,p_0,0) \cdot \exp(p_0\theta') \cdot [1 - u(x' - 1)] + \bar{C}'(1,p_0,p') \cdot \delta(x' - 1) \qquad (3\text{-}10)$$

This ordinary differential equation is easily integrated by using x' Laplace transforms, give $\bar{C}'(x',p_0,p')$:

$$\bar{C}' = \frac{C'(1,p_0,0)}{p' + p_0}\left\{\exp(p_0\theta') \cdot [1 - u(x' - 1)] + \exp(-p'\theta') \cdot u(x' - 1) - \exp\left[-(p_0 l + p'x')\frac{(1 + k_A)}{v}\right]\right\} + \bar{C}'(1,p_0,p')\exp(-p'\theta') \cdot u(x' - 1) \qquad (3\text{-}11)$$

At the detector, i.e., for x' = l' + 1, u(x' − 1) becomes u(l') = 1, for l' > 0, and θ' becomes $(1 + k_A)l'/v = t_R'$, i.e., the ideal retention time of component A in the column section l'. Then Eq. (3-11) becomes

$$\bar{C}' = \bar{C}'(1,p_0,p')\exp(-p't_R') + \frac{C'(1,p_0,0)}{p' + p_0}\left\{\exp(-p't_R') - \exp[-p'(t_R' + t_R)] \cdot \exp(-p_0 t_R)\right\} \qquad (3\text{-}12)$$

The break-through curve of A at the end D_2 of the column is now obtained by taking the inverse transforms, first with respect to p' and then with respect to p_0. The result is

$$c' = c_1'(l, t_0 + \tau') \cdot u(\tau') + c_2'(l, t_0 - \tau') \cdot [u(\tau') - u(\tau' - t_R)] \cdot u(t_0 - \tau') \tag{3-13}$$

where

$$\tau' = t' - t_R' \tag{3-14}$$

and $c_1'(l, t_0, t' - t_R')$ is written $c_1'(l, t_0 + \tau')$, since the time t' is a continuation of t_0.

Equation (3-13) is the simplest chromatographic sampling equation. Its behavior for various values of τ' is interesting and is illustrated diagrammatically in Fig. 3.5. For $\tau' < 0$, i.e., for $t' < t_R'$, $c' = 0$ and no signal is recorded by the detector until the retention time t_R' is reached. Then $u(\tau') = 1$, and the chromatographic signal rises abruptly to $c_1' + c_2'$. It falls again to c_1' when $\tau' \geq t_R$, i.e., when $t' \geq t_R' + t_R$, because the square function in brackets [ ] becomes zero. The $u(t_0 - \tau')$ factor remains at unity in the above interval, since the flow was reversed at $t_0 > t_R$. Thus the concentration of A at x = l', $c(l', t_0)$, or at x' = l, $c'(l, t_0)$ owing to the rate process at this position, is shifted in time on reversing the flow direction. This time shift takes place in two opposite directions. One, which occurs forward to $c_1'(l, t_0 + \tau')$, starts at $t' = t_R'$ and continues uninterrupted. It is nothing else than the continuation of Eq. (3-5) at the other end of the column. The second shift $c_2'(l, t_0 - \tau')$ occurs backward and is barred in the interval $0 \leq \tau' \leq t_R$ or $t_R' \leq t' \leq t_R' + t_R$. Therefore, it starts with the concentration $c_2'(l, t_0)$ and ends with that of a preceding time, i.e., $c_2'(l, t_0 - t_R)$. This extra signal (R-peak; see Fig.

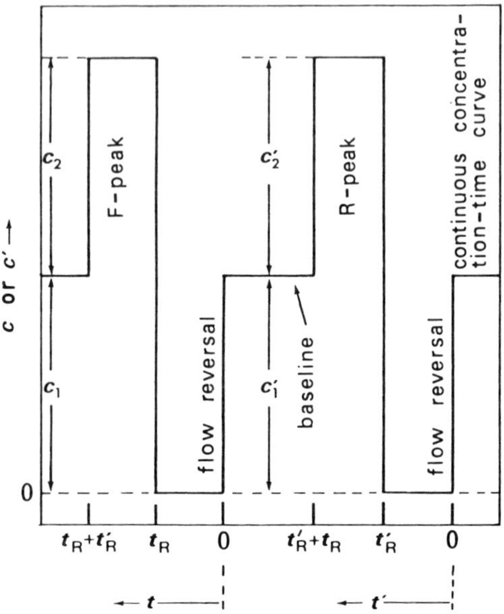

Fig. 3.5 Elution curves predicted by Eqs. (3-13) and (3-21).

3.5) adds to the forward shift, and constitutes the simplest form of a sample peak for the rate process under investigation.

C. Second F-Interval

After a certain time, t', of backward flow, the carrier gas is again turned to the direction F, the time from this moment being denoted by t. The distance coordinate is changed from x' back to x, according to Eq. (3-6), and the concentration c(x,t) is again described by Eq. (3-1) with t substituted for t_0:

$$(1 + k_A) \frac{\partial c}{\partial t} = - v \frac{\partial c}{\partial x} + v c(l',t)\delta(x - l') \qquad (3\text{-}15)$$

This is integrated by taking successive Laplace transformations with respect to t_0, t', and t. The initial condition at t = 0 (in the form of its t_0 and t' double transform) is obtained from

Eq. (3-11) by changing x' to l' + l − x, θ' to − θ, and u(x' − l) to 1 − u(x − l'), according to Eqs. (3-6) and (3-9), with the result:

$$\bar{C}(x,p_0,p',0) = \frac{C(l',p_0,0)}{p' + p_0}\Bigg\{\exp(-p_0\,\theta)\cdot u(x - l')$$

$$+ \exp(p'\theta)\cdot[1 - u(x - l')]$$

$$- \exp\left[-\frac{(p_0 + p')(1 + k_A)l}{v} + p'\,\theta\right]\Bigg\}$$

$$+ \bar{C}(l',p_0,p')\cdot\exp(p'\theta)\cdot[1 - u(x - l')]$$

(3-16)

The result of the procedure outlined above is an ordinary differential equation in x, which can easily be integrated by using x Laplace transforms, giving $\bar{\bar{C}}(x,p_0,p',p)$:

$$\bar{\bar{C}} = \frac{C(l',p_0,0)}{p' + p_0}\Bigg\{\frac{\exp(-p_0\,\theta) - \exp(-p\,\theta)}{p - p_0}$$

$$\cdot\, u(x - l') - \frac{\exp(p'\theta) - \exp(-p\,\theta)}{p + p'}$$

$$\cdot\, u(x - l') + \frac{\exp(p'x/v_A) - \exp(-px/v_A)}{p + p'}$$

$$\cdot\,\exp\left(-\frac{p'l'}{v_A}\right)\left[1 - \exp\left(-\frac{p_0 l}{v_A}\right)\cdot\exp\left(-\frac{p'l}{v_A}\right)\right]\Bigg\}$$

$$+ \frac{\bar{C}(l',p_0,p')}{p + p'}\Bigg\{\exp(p'\theta) - \exp\left(-\frac{p'l' + px}{v_A}\right)$$

$$- [\exp(p'\theta) - \exp(-p\,\theta)]\cdot u(x - l')\Bigg\}$$

$$+ \bar{\bar{C}}(l',p_0,p',p)\exp(-p\theta)\cdot u(x - l') \qquad (3\text{-}17)$$

where

$$v_A = \frac{v}{1 + k_A} \tag{3-18}$$

At the detector Eq. (3-17) is considerably simplified, since there $x = l' + l$, $u(x - l') = 1$ and $\theta = (1 + k_A)l/v = t_R$. Then, performing inverse Laplace transformations with respect to p, p' and p_0 in succession, one finds the final equation giving the value of c at the detector:

$$\begin{aligned} c = {} & c_1(l', t_0 + t' + \tau) \cdot u(\tau) + c_2(l', t_0 + t' - \tau) \\ & \cdot [1 - u(\tau - t')] \cdot [u(\tau) - u(\tau - t_R')] \\ & + c_3(l', t_0 - t' + \tau) \cdot u(t_0 + \tau - t') \\ & \cdot \{u(t - t')[1 - u(\tau - t_R')] \\ & - u(\tau - t')[u(\tau) - u(\tau - t_R')]\} \end{aligned} \tag{3-19}$$

where

$$\tau = t - t_R \tag{3-20}$$

Equation (3-19) is the desired <u>general chromatographic sampling equation</u>. On the right-hand side, it consists of three concentration terms, denoted by c_1, c_2, and c_3. They all refer to $x = l'$ but to different values of the time variable; i.e., $t_0 + t' + \tau$, $t_0 + t' - \tau$, and $t_0 - t' + \tau$, respectively. Each of the concentration terms is multiplied by a combination of unit step functions, so that it appears in a certain time interval and vanishes in all others. There are various possibilities depending on the relative values of t_0, t', t, t_R, and t_R'.

1. Case Where $t' > t_R + t'_R$

This condition reduces Eq. (3-19) to

$$c = c_1(l', t_0 + t' + \tau) \cdot u(\tau) + c_2(l', t_0 + t' - \tau) \cdot [u(\tau) - u(\tau - t'_R)] \quad (3\text{-}21)$$

since the term c_3 becomes zero for all values of t, and the factor $[1 - u(\tau - t')]$ of the c_2 term remains at unity in the interval defined by the other factor $[u(\tau) - u(\tau - t'_R)]$. Equation (3-21) is the F-direction equivalent of Eq. (3-13) and its behavior is analogous to that. It is also depicted in Fig. 3.5 and predicts that $c = 0$ for $\tau < 0$, i.e., $t < t_R$, and at $\tau \geq 0$ or $t \geq t_R$ two functions are recorded as a sum $c_1 + c_2$. The first is $c(l', t_0 + t')$ shifted forward by τ. This continuous uninterrupted as the first term of Eq. (3-21) shows. In the other function, c_2, the total time $t_0 + t'$ is shifted backward by τ, and this function vanishes when $\tau \geq t'_R$ or $t \geq t_R + t'_R$. Thus, a sample peak (F-peak; see Fig. 3.5) is predicted in the interval $t_R \leq t \leq t_R + t'_R$, positioned on top of the otherwise continuing chromatographic curve.

Repeating the reversal of the carrier gas flow in the direction R for a second time, then in the direction F for a third time, and so on, keeping the time between any two successive reversals greater than the total retention time $t^0_R + t'_R$, two series of peaks are produced. The R-peaks are described by Eq. (3-13), whereas the F-peaks are given by Eq. (3-21). Since the definitions forward and reverse are arbitrary, and the two above equations have the same form, one of them, say Eq. (3-13), suffices to describe both kinds of peaks. In that case, t_0 is taken to represent the total time passed from the beginning to the last reversal of gas flow, τ', the time passed from the last reversal of the flow diminished by the retention time of A in the new flow direction, and t_R, the retention time in the opposite direction.

2. Case Where $t' < t_R + t'_R$

This condition retains all three terms of the sampling equation, Eq. (3-19). The first term, c_1, appears at $t = t_R$ ($\tau = 0$) and continues uninterrupted. The second term, c_2, appears again at $t = t_R$, but it is cut down either at $t = t_R + t'$($\tau = t'$) or at $t = t_R + t'_R$ ($\tau = t'_R$), whichever comes first. This, of course, depends on whether $t' < t'_R$ or $t' > t'_R$.

The third term, c_3, appears at $t = t'$ and vanishes at the same t value as the second term. Thus for $\tau \geq t'$ or $\tau \geq t'_R$, only c_1 remains as an <u>ending baseline</u>. Before that there are two possibilities: 1. $t' < t_R$, and 2. $t' > t_R$. In case 1. the c_3 term appears first and then "sitting" on it the sum $c_1 + c_2$. This is shown in Fig. 3.6A. In case 2., the sum $c_1 + c_2$ appears first and then on it c_3 (Fig. 3.6B).

If t' is small enough, the c_1 and c_3 terms will differ little in their time argument, and the situation depicted in Fig. 3.6A will have the appearance of a relatively narrow peak of two terms, with baseline the remaining term (c_3 as the starting and c_1 as the ending baseline). The smaller t' is, the narrower this <u>sample peak</u> becomes, as its width is clearly equal to t'. The situation in Fig. 3.6B is less favorable for measurements because a single term appears to "sit" on a starting baseline consisting of two terms ($c_1 + c_2$) and on an ending baseline of only c_1.

In Fig. 3.7 a combination of Figs. 3.5 and 3.6 is drawn, showing the two most important limiting cases of Eq. (3-19); i.e., $t' > t_R + t'_R$ and $t' < t_R$, $t' < t'_R$. Comparison of the sample peaks of this figure with those of Fig. 3.3 found experimentally shows that the elution curves predicted by Eq.

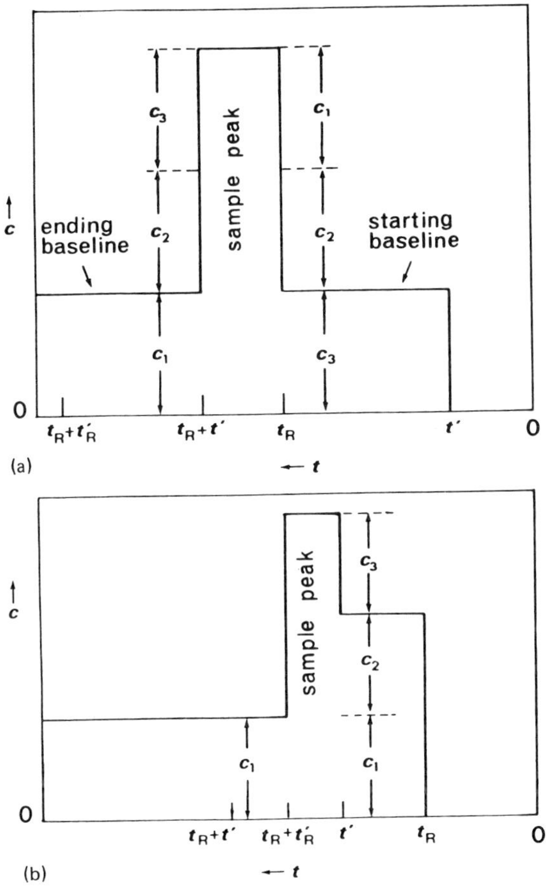

Fig. 3.6 Elution curves predicted by Eq. (3-19) for $t' < t_R + t'_R$: (a) when $t' < t_R$ and $t' < t'_R$, and (b) when $t' > t_R$ and $t' > t'_R$.

(3-19) are confirmed experimentally, apart from the fact that the peaks actually found are not square owing to nonideality (e.g., axial diffusion in column sections l and l').

In summary, the chromatographic sampling of a process taking place at a certain point within a chromatographic column is effected by reversing the flow direction of the carrier

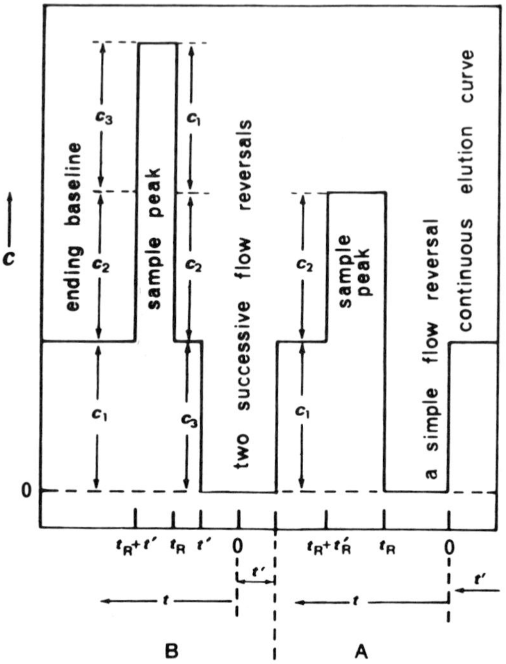

Fig. 3.7 Elution curves predicted by Eq. (3-19): (A) when $t' > t_R + t'_R$, and (B) when $t' < t_R$ and $t' < t'_R$.

gas. This creates sample peaks in the elution curve. When the time elapsing between any two successive reversals of the flow is greater than the total retention time, $t_R + t'_R$, of the component examined on column $l + l'$, a sample peak follows <u>every</u> reversal, as shown diagrammatically in Fig. 3.5. This is described mathematically by Eq. (3-13) or (3-21). But when the time passing between two successive reversals is less than $t_R + t'_R$, only one sample peak appears after the <u>two</u> successive reversals, as predicted by the general Eq. (3-19). This peak is either symmetrical when $t' < t_R$ (Fig. 3.6A) or unsymmetrical when $t' > t_R$ (Fig. 3.6B).

V. CALCULATION OF RATE COEFFICIENTS AND EQUILIBRIUM CONSTANTS FROM THE SAMPLE PEAKS

If the concentrations c_1, c_2, c_3 of Eq. (3-19), pertaining to $x = l'$, are due to a rate process or an equilibrium state occurring inside the sampling cell, the rate coefficient or the equilibrium constant of that process can be determined from the sample peaks created by applying the reversed-flow technique. For this purpose, either the area under the sample peaks, f, or their height, h, can be used. Theoretically the area f can most easily be calculated from the equations preceding Eq. (3-19) and its limiting case, Eq. (3-13), i.e., Eq. (3-17) and Eq. (3-12), respectively. As an example, let us calculate the area under the R-peak of Fig. 3.5. According to a well-known relation:

$$\underset{t_0}{\mathscr{L}} \quad f = \int_0^\infty C'dV = \dot{V}\int_0^\infty C'dt' = \dot{V} \lim_{(p'\to 0)} (\bar{C}') \tag{3-22}$$

where in place of $\bar{C}'$ we use the second term of Eq. (3-12, since it is this term giving rise to the sample R-peak, the first term being responsible for the ending baseline. The result is

$$\underset{t_0}{\mathscr{L}} \quad f = \frac{\dot{V}C'(l,p_0)}{p_0} [1 - \exp(-p_0t_R)] \tag{3-23}$$

The area under the peak is found by simply taking the p_0 inverse transform of this expression.

The height, h, of the sample peaks above the continuous chromatographic signal (baseline) is proportional to either one or two concentration terms of Eq. (3-19) (cf. Fig. 3.7). For the case of the sample peak of Fig. 3.7A, this height is proportional to the term c_2 of Eq. (3-21). It can be measured from the baseline to either of the discontinuities of the peak corresponding to $t = t_R$ ($\tau = 0$) or $t = t_R + t'_R$ ($\tau = t'_R$). Thus using Eq. (3-21):

$$h_{\tau=0} = c_2(l', t_0 + t') \qquad (3\text{-}24)$$

$$h_{\tau=t'_R} = c_2(l', t_0 + t' - t'_R) \qquad (3\text{-}25)$$

For the case of the sample peak of Fig. 3.7B, its maximum is taken to correspond to its middle time, i.e., to $t = t_R + t'/2$ or $\tau = t'/2$. The height, h, of this maximum, measured from the ending baseline, gives the sum $c_2 + c_3$. From Eq. (3-19), by putting $\tau = t'/2$ in these two terms, one obtains

$$h = c_2\left(l', t_0 + \frac{t'}{2}\right) + c_3\left(l', t_0 - \frac{t'}{2}\right) \qquad (3\text{-}26)$$

The times in the two terms on the right-hand side differ only by t', and if this is small, both terms can be taken at a mean time, t_0. Then the above relation becomes simply

$$h \simeq 2c(l', t_0) \qquad (3\text{-}27)$$

Thus sample peaks like that of Fig. 3.7B are approximately double in height than that of Fig. 3.7A, and consequently the use of such peaks increases the precision of the method.

By following either the area f under the sample peaks or their height, h, as a function of the corresponding time, as indicated in Eqs. (3-23) to (3-27), the rate coefficient or the equilibrium constant of the process responsible for the various concentrations is determined, as described in the following chapters.

LIST OF SYMBOLS

a	Volume of gas phase per unit length of column, or cross-sectional area of void space
c,c'	Concentrations of a solute vapor in the chromato-

	graphic column with the carrier gas flowing in direction F or R, respectively
C, C'	Laplace transforms of c and c' with respect to t_0
$\bar{C}, \bar{C}'$	Double Laplace transforms of c and c' with respect to t_0 and t'
$\bar{\bar{C}}$	Triple Laplace transform of c with respect to t_0, t', and t
f	Area under the curve of R- and F-peaks
h	Height above the baseline defined by Eqs. (3-24) to (3-27)
k_A	Partition ratio
l, l'	Lengths of two sections of the chromatographic column (Fig. 3.4)
p_0, p', p	Transform parameters with respect to t_0, t', and t respectively
t_0	Time measured from the beginning to the last backward reversal of gas flow
t'	Time interval of backward flow of carrier gas
t	Time measured from the last restoration to the forward direction of the gas flow
t_{tot}	Sum of times t_0 and t'
t_m, t'_M	Gas hold-up time of the empty column section l or l', respectively
t_R, t'_R	Ideal retention time on the filled column section l or l', respectively
v	Linear velocity of carrier gas in interparticle space of the sampling column
v_A	Velocity defined by Eq. (18)
$\dot{V}$	Volume flowrate of carrier gas
x, x'	Distance coordinates defined in Fig. 3.4
θ, θ'	Time parameter defined by Eqs. (3-3) and (3-9), respectively
τ, τ'	Time defined by Eqs. (3-20) and (3-14), respectively

REFERENCES

1. N. A. Katsanos and G. Karaiskakis, Advan. Chromatogr., 24:125 (1984).
2. N. A. Katsanos and I. Georgiadou, J. Chem. Soc. Chem. Commun., 242 (1980).
3. N. A. Katsanos, J. Chem. Soc. Faraday Trans. I., 78:1051 (1982).
4. G. Karaiskakis, N. A. Katsanos, I. Georgiadou, and A. Lycourghiotis, J. Chem. Soc. Faraday Trans. I., 78:2017 (1982).
5. M. Kontinopoulos, G. Karaiskakis, and N. A. Katsanos, J. Chem. Soc. Faraday Trans. I., 78:3379 (1982).
6. N. A. Katsanos, ACHEMA 82 International Meeting of Chemical Engineering, Frankfurt -am-Main, 1982.
7. N. A. Katsanos and M. Kontinopoulos, J. Chem. Soc. Faraday Trans. I., 81:951 (1985).
8. G. Karaiskakis, N. A. Katsanos, and A. Lycourghiotis, Can. J. Chem., 61:1853 (1985).
9. N. A. Katsanos, G. Karaiskakis, and A. Niotis, Proc. 8th Inter. Congress on Catalysis, West Berlin, 1984 (Dechema, Verlag chemie), Vol. III, p. 143
10. G. Karaiskakis and N. A. Katsanos, Proc. 3rd Mediterranean Congress on Chemical Engineering, Barcelona, 1984, p. 68.
11. N. A. Katsanos, G. Karaiskakis, and A. Niotis, J. Catal., 94:376 (1985).
12. E. Dalas, N. A. katsanos, and G. Karaiskakis, J. Chem. Soc. Faradya Trans. I., 82:2897 (1986).
13. G. karaiskakis, N. A. Katsanos, and A. Niotis, J. Chromatogr., 245:21 (1982).
14. N. A. Katsanos and G. Karaiskakis, J. Chromatogr., 237:1 (1982).
15. N. A. Katsanos and G. Karaiskakis, J. Chromatogr., 254:15 (1983).
16. G. Karaiskakis, N. A. Katsanos, and A. Niotis, Chromatographia, 17:310 (1983).
17. N. A. Katsanos, G. Karaiskakis, and A. Niotis, 1983 World Chromatography/Spectroscopy Conference, London, 1983.

18. G. Karaiskakis, A. Lycourghiotis, and N. A. Katsanos, Chromatographia, 15:351 (1982).

19. G. Karaiskakis and N. A. Katsanos, J. Phys. Chem., 88: 3674 (1984).

20. G. Karaiskakis, A. Niotis, and N. A. Katsanos, J. Chromatogr. Sci., 22:554 (1984).

21. N. A. Katsanos, G. Karaiskakis, and P. Agathonos, J. Chromatogr., 349:339 (1985).

22. E. Dalas, G. Karaiskakis, N. A. Katsanos, and A. Gounaris, J. Chromatogr., 348:339 (1985).

23. G. Karaiskakis, P. Agathonos, A. Niotis, and N. A. Katsanos, J. Chromatogr., 364:79 (1986).

24. G. Karaiskakis, J. Chromatogr. Sci., 23:360 (1985).

25. A. Niotis, N. A. Katsanos, G. Karaiskakis, and M. Kotinopoulos, Chromatographia, 23:447 (1987).

4

Reversed-Flow with an Empty Diffusion Column

I. Gas Diffusion Coefficients in Binary Mixtures

A. Experimental Details
B. Theory
C. Some Experimental Results

II. Diffusion Coefficients in Multicomponent Gas Mixtures

III. Relative Molar Responses, Collision Diameters, and Critical Volumes of Gases

A. Theory
B. Experimental Details
C. Some Typical Results

IV. Lennard-Jones Parameters

List of Symbols

References

I. GAS DIFFUSION COEFFICIENTS IN BINARY MIXTURES

The simplest conceivable case of reversed-flow gas chromatography (RFGC) is to have the sections l' and l shown in Fig. 3.1 entirely empty of any chromatographic material, the rate process at $x = l'$ being the slow diffusion of a gaseous solute A into the carrier gas B from the diffusion column L connected perpendicularly to column l' + l at x = l'. It permits the measurement of the mutual diffusion coefficient of A and B, this being an appropriate rate coefficient for the process (1). The component B enters at D_2 and meets the detector at D_1, or vice versa, flowing continuously through the sampling column l' + l, and filling also the diffusion column L. At the closed end of the latter the solute A is introduced (by means of a syringe) as a gas or vapor in the form of a small volume pulse or a bigger volume so that it fills a considerable part or even the entire length, L.

As mentioned in Chapter 3, section II, the diffusion column L is relatively short (about 30-100 cm) and can be straight or coiled and accommodate inside the chromatograph's oven with the closed end at the injector position.

A. Experimental Details

1. Apparatus

The experimental setup for the application of the reversed-flow method in this case is particularly simple, and needs only a slight modification of a common gas chromatograph, as shown diagrammatically in Fig. 4.1. Any kind of gas chromatographic detector can be used, although a high-sensitivity device, like a flame ionization detector (FID), is to be preferred. The reversing of the flow is effected, as previously described in Chapter 3, section II, by means of valve S (four-port or six-port).

2. Procedure

While carrier gas B is flowing in direction F (Fig. 4.1; valve S in position indicated by the solid lines), a small amount of solute A (usually 0.5-10 cm^3 of gas at atmospheric pressure) is injected into the diffusion column, L. After a certain time, during which no signal is noted, an asymmetric concentration-time curve of A is recorded, usually decaying slowly. At a certain known time (from the moment of injection) the direction of the carrier gas is

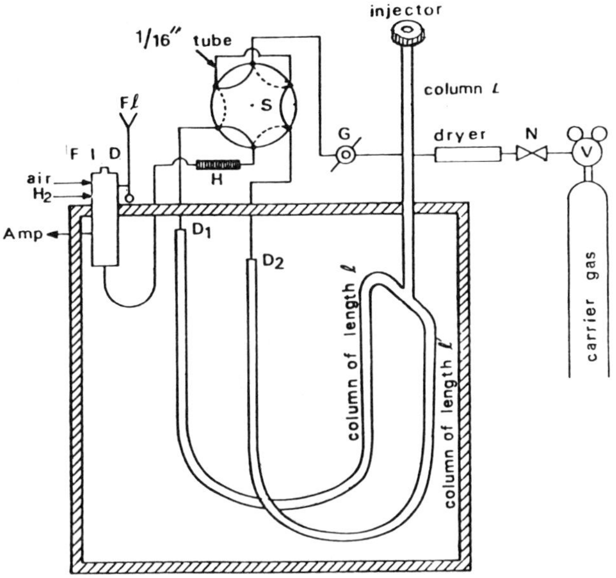

Fig. 4.1 Schematic arrangement of the experimental setup for measuring gas diffusion coefficients: V, two-stage reducing valve and pressure regulator; N, needle valve; G, gas flow controller for minimizing variations in the gas flowrate; S, six-port gas sampling valve; H, restrictor; Fl, bubble flowmeter; and Amp, signal to amplifier. (From Ref. 1, used with permission.)

reversed by switching valve S to the other position (dashed lines). This reversal is repeated several times, the interval between successive reversals depending on whether the condition $t' > t_R + t'_R$ or $t' < t_R + t'_R$ is adopted (see Chapter 3, section IV.C and Fig. 3.7). A whole series of sample peaks is thus obtained.

The pressure drop along column $l' + l$ is negligible, and the diffusion coefficients are considered to have been determined at the pressure measured near the injection point by means of a suitable manometer.

The temperature of the diffusion column L can be regulated, if necessary, by conventional methods. The simplest way is to coil this column and place it inside the chromatographic oven, with its closed end at the injector position of the chromatograph. This arrangement is particularly suitable to study the temperature variation of gas diffusion coefficients (2).

B. Theory

In order to apply the chromatographic sampling equation, Eq. (3-19), and all other relations derived from it, the concentration $c(l',t_0)$ of the solute A is required, at the junction $x = l'$ of the sampling and the diffusion column, as a function of time t_0 (see Fig. 4.2). This is determined by the diffusion equation (Fick's second law):

$$\frac{\partial c_z}{\partial t_0} = D \frac{\partial^2 c_z}{\partial z^2} \tag{4-1}$$

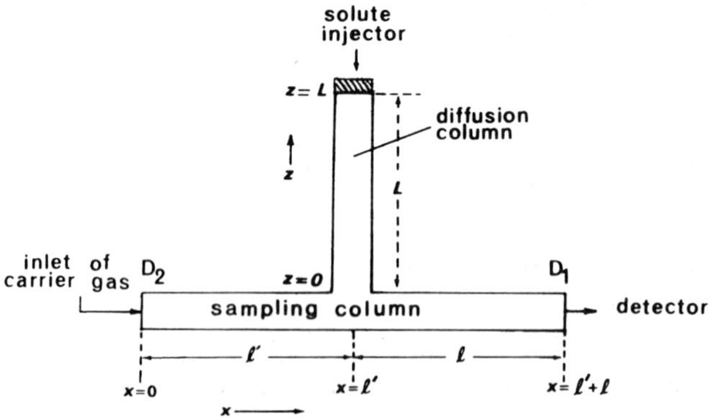

Fig. 4.2 Arrangement of a diffusion column L and a sampling column l' + l for diffusion coefficient measurements showing the relevant coordinates x and z. (Adapted from Ref. 1, with permission.)

where $c_z = c_z(z,t_0)$ is the concentration of the solute vapor A in the diffusion column L, and D the diffusion coefficient of A into the carrier gas B. This partial differential equation can be solved using the method of Laplace transforms under given initial and boundary conditions. Since there is no flux of A across the boundary at z = L:

$$\left(\frac{\partial c_z}{\partial z}\right)_{z=L} = 0 \tag{4-2}$$

whereas at the other boundary z = 0:

$$c_z(0,t_0) = c(l',t_0) \tag{4-3}$$

and

$$D\left(\frac{\partial c_z}{\partial z}\right)_{z=0} = vc(l',t_0) \tag{4-4}$$

where v is the linear velocity of the carrier gas B.

The initial condition depends on the mode of introducing the solute A into column L through the point z = L. Using a gas tight syringe, A can be introduced as:

1. A rapid pulse of 0.2-0.5 cm^3 of gas or vapor at atmospheric pressure.
2. A slow injection of a bigger volume, so as to fill a considerable part of L with A.
3. A much bigger volume of gas (5-10 cm^3 at atmospheric pressure), so as to fill initially the entire length of the diffusion column L with an approximately uniform concentration in A.

The solution of the differential equation (4-1) will now be found separately for each of these three initial conditions.

1. Solution with a Pulse Initial Condition

This condition is described by a Dirac delta function:

$$c_z(z,0) = \frac{m}{a}\,\delta(z - L) \tag{4-5}$$

where m is the amount of solute injected and a the cross-sectional area of void space in the column L.

Laplace transformation with respect to time t_0 (trnasform parameter p_0) of Eq. (4-1) as well as of the boundary conditions (4-2), (4-3), and (4-4), under the initial condition (4-5), gives

$$p_0 C_z - \frac{m}{a}\,\delta(z - L) = D\,\frac{d^2C_z}{dz^2} \tag{4-6}$$

$$\left(\frac{dC_z}{dz}\right)_{z=L} = 0 \tag{4-7}$$

$$C_z(0,p_0) = C(l',p_0) \tag{4-8}$$

$$D\left(\frac{dC_z}{dz}\right)_{z=0} = vC(l',p_0) \tag{4-9}$$

where we used capital letters C_z and C to denote the t_0 Laplace transformed functions of c_z and c, respectively.

Eq. (4-6) written in the form

$$\frac{d^2C_z}{dz^2} - q^2C_z = -\frac{m}{aD}\,\delta(z - L) \tag{4-10}$$

where

$$q^2 = \frac{p_0}{D} \tag{4-11}$$

is an ordinary linear second-order equation, which can be integrated by using z Laplace transformation (transform parameter s):

$$s^2\bar{C}_z - sC_z(0) - C_z'(0) - q^2\bar{C}_z = -\frac{m}{aD}\exp(-sL) \qquad (4\text{-}12)$$

where $\bar{C}_z$ is the double Laplace transform of c_z with respect to t_0 and z, $C_z(0)$ and $C_z'(0)$ being the t_0 transforms of the concentration c_z and its first z derivative, respectively, at z = 0, i.e., $C_z(0,p_0)$ and $(dCz/dz)_{z=0}$, respectively. Inverse Laplace transformation of Eq. (4-12) with respect to the parameter s gives

$$C_z = C_z(0)\cosh qz + \frac{C_z'(0)}{q}\sinh qz - \frac{m}{aDq}\sinh q(z - L) \cdot u(z - L) \qquad (4\text{-}13)$$

where u(z − L) is the unit step function with values 0 for $z < L$ and 1 for $z \geq L$. Finally, using the transformed boundary conditions (4-7), (4-8), and (4-9) to replace $C_z(0)$ by $C(l',p_0)$ according to (4-8), $C_z'(0)$ by $vC(l',p_0)/D$ according to (4-9), differentiating the result with respect to z and setting the derivative at z = L equal to 0 according to (4-7), one obtains

$$C(l',p_0) = \frac{m}{aDq} \cdot \frac{1}{\sinh qL + (v/Dq)\cosh qL} \qquad (4\text{-}14)$$

Inverse Laplace transformation of this equation to find $c(l',t_0)$ is difficult. It can be achieved by using certain approximations, the first of which is to omit sinh qL in the denominator. This is based on the fact that $\sinh qL \leq \cosh qL$ and that for high enough flowrates, $v/Dq \gg 1$. Eq. (4-14) then reduces to

$$C(l',p_0) = \frac{m}{\dot{V}}\operatorname{sech} qL \qquad (4\text{-}15)$$

where $\dot{V}$ is the volumetric flowrate in the sampling column l' + l. Taking now the inverse transform of (4-15) in the form of an elliptic theta function (3), we obtain

$$c(l',t_0) = -\frac{N_1}{t_0^{3/2}} \sum_{n=-\infty}^{\infty} (-1)^n \exp\left[-\left(n - \frac{1}{2}\right)^2 \frac{L^2}{Dt_0}\right]\left(n - \frac{1}{2}\right) \tag{4-16}$$

where

$$N_1 = \frac{mL}{\dot{V}(\pi D)^{1/2}} \tag{4-17}$$

If we write explicitly the first four terms around n = 0 in the summation, i.e., with n = −1, 0, 1, 2, we have

$$c(l',t_0) = \frac{N_1}{t_0^{3/2}}\left[\exp\frac{-L^2}{4Dt_0} - 3\exp\left(\frac{-9L^2}{4Dt_0}\right) + \cdots\right] \tag{4-18}$$

By using long enough diffusion columns (say with L = 80 cm), and taking measurements at small times we can ignore the second term inside the brackets [ ], since it contains an exponent nine times bigger than that of the first term. Finally, we are left with the equation

$$c(l',t_0) = \frac{N_1 \exp(-L^2/4Dt_0)}{t_0^{3/2}} \tag{4-19}$$

Using now Eq. (4-19) in place of the functions c_1, c_2, and c_3 of Eq. (3-19), we can obtain c as an explicit function of the time, thus describing analytically the concentration-time curve, as recorded by the detector system. For example, c_1 will be given by the expression

$$c_1 = \frac{N_1 \exp[-L^2/4D(t_0 + t' + \tau)]}{(t_0 + t' + \tau)^{3/2}} \tag{4-20}$$

where

$$\tau = t - t_M \tag{4-21}$$

according to Eq. (3-20), since column 1 is empty and the gas hold-up time in it, t_M, must be substituted for t_R.

Also, Eq. (3-25) becomes

$$h_{\tau=t'_M} = \frac{N_1 \exp[-L^2/4D(t_0 + t' - t'_M)]}{(t_0 + t' - t'_M)^{3/2}} \tag{4-22}$$

where t'_M has been substituted for t'_R. Eq. (3-27) reads

$$h = \frac{2N_1 \exp(-L^2/4Dt_0)}{t_0^{3/2}} \tag{4-23}$$

The experimental verification of Eq. (4-22) is made by plotting $\ln[h(t_0 + t' - t'_M)^{3/2}]$ versus $1/(t_0 + t' - t'_M)$, where h is the height of the peaks measured from the continuous chromatographic signal to the maximum of the peak before returning to the baseline (cf. sample peak A in Fig. 3.3). Similarly, Eq. (4-23) is verified by plotting $\ln(ht_0^{3/2})$ against $1/t_0$, the height h being measured from the ending baseline to the peak maximum (cf. sample peak B in Fig. 3.3). In both kinds of plot, the diffusion coefficient is easily computed from the slope of the straight lines obtained and the known value of L.

The basic Eq. (4-19) was obtained as an approximation to Eq. (4-16), which was the result of the inverse Laplace transformation of Eq. (4-15) using the elliptic theta function $\theta_1(\varepsilon/2 \mid (t_0D)/L^2)$ in the form of an infinite series of exponentials of $1/t_0$. However, theta functions are also expressed (3) as infinite series of exponentials of t_0, and using the expression for θ_1, the inverse transform of (4-15) comes out as

$$c(l',t_0) = N_2 \sum_{n=0}^{\infty} (-1)^n(2n+1) \exp\left[-\left(n+\frac{1}{2}\right)^2 \frac{\pi^2 D t_0}{L^2}\right] \tag{4-24}$$

where

$$N_2 = \frac{\pi m D}{\dot{V} L^2} \tag{4-25}$$

If the first two terms in the summation are written explicitly, this equation reads:

$$c(l', t_0) = N_2\left[\exp\left(\frac{-\pi^2 D t_0}{4L^2}\right) - 3\exp\left(\frac{-9\pi^2 D t_0}{4L^2}\right) + \cdots\right] \tag{4-26}$$

This is another convenient expression for plotting experimental data of diffusion, particularly with short column length L (say 30–50 cm) and long times, since under these conditions, a direct plot of ln c versus t_0 becomes quickly linear, after the maximum owing to the last term on the right-hand side of Eq. (4-26). Taking, therefore, only the first exponential and combining it with the equations of Chapter 3, section V, say Eq. (3-27), one obtains

$$h = 2N_2 \exp\left(\frac{-\pi^2 D t_0}{4L^2}\right) \tag{4-27}$$

Thus the diffusion coefficient D can be found from the linear plot of ln h versus t_0.

The choice between Eq. (4-19), which is a good approximation for long L and small t_0, and Eq. (4-27), which is to be preferred with a short L and a long t_0, must be a matter of preliminary experimentation with a gas of known D value, possibly close to that expected for the unknown D.

2. Solution with the Solute Partly Filling Column L Initially

Instead of injecting the solute A as a small volume pulse, described by the initial condition (4-5), we can introduce more slowly into the diffusion column a bigger volume of A, so as to fill a part of the total length, L, with it. The initial condition now can be described by the equation:

$$c_z(z,0) = c_0 u(z - b) \tag{4-28}$$

where c_0 is the initial concentration created by the injection of A and $u(z - b)$ the unit step function. This is 0 for $z < b$ and 1 for $z \geq b$, the b denoting the length which is empty.

Laplace transformation with respect to time t_0 of Eq. (4-1) now gives (after some rearrangement)

$$\frac{d^2 C_z}{dz^2} - q^2 C_z = -\frac{c_0}{D} u(z - b) \tag{4-29}$$

where q^2 is given again by Eq. (4-11). As before, Eq. (4-29) can be integrated by using z Laplace transformation and evaluating $C_z(0)$ and $C'_z(0)$ by means of the boundary conditions (4-8) and (4-9), respectively. Then, differentiating with respect to z and setting the derivative at $z = L$ equal to 0, according to (4-7), we obtain

$$C(l',p_0) = \frac{c_0}{Dq^2} \cdot \frac{\sinh q(L - b)}{\sinh qL + (v/Dq)\cosh qL} \tag{4-30}$$

The inverse transformation with respect to p_0 to find $c(l',t_0)$ is effected as before by omitting sinh qL in the denominator of (4-30), because $v/D_q \gg 1$, resulting in

$$C(l',p_0) = \frac{c_0}{vq} \cdot \frac{\sinh q(L - b)}{\cosh qL} \tag{4-31}$$

The inverse Laplace transform of this equation in the form of an elliptic theta function (3) is

$$c(l',t_0) = \frac{N_3}{t_0^{1/2}} \sum_{n=-\infty}^{\infty} (-1)^n \exp\left[-\left(\frac{L-b}{2L} + n - \frac{1}{2}\right)^2 \frac{L^2}{Dt_0}\right]$$

$$= N_4 \sum_{n=0}^{\infty} (-1)^n \exp\left[-\pi^2 D\left(n + \frac{1}{2}\right)^2 \frac{t_0}{L^2}\right]$$

$$\cdot \sin\left[(2n+1)\pi \cdot \frac{L-b}{2L}\right] \qquad (4\text{-}32)$$

where

$$N_3 = \frac{c_0 D^{1/2}}{v\pi^{1/2}} \qquad N_4 = \frac{2c_0 D}{vL} \qquad (4\text{-}33)$$

Eq. (4-32) is analogous to Eqs. (4-16) and (4-24). Writing explicitly the first four terms around n = 0 (−1, 0, 1, 2) in the first summation, and the first two terms (n = 0, 1) in the second summation, Eq. (4-32) reads

$$c(l',t_0) = \frac{N_3}{t_0^{1/2}} \left\{\exp\left[-\frac{(b/2)^2}{Dt_0}\right] - \exp\left[-\frac{(L-b/2)^2}{Dt_0}\right]\right.$$

$$\left. - \exp\left[-\frac{(L+b/2)^2}{Dt_0}\right] + \exp\left[-\frac{(2L-b/2)^2}{Dt_0}\right]\right\}$$

$$= N_4\left\{\exp\left(\frac{-\pi^2 Dt_0}{4L^2}\right)\sin\left[\pi\frac{L-b}{2L}\right]\right.$$

$$\left. - \exp\left(\frac{-9\pi^2 Dt_0}{4L^2}\right)\sin\left[3\pi\frac{L-b}{2L}\right]\right\} \qquad (4\text{-}34)$$

One can try various values of b in this expression, leading to slightly different final equations. Take, for instance, $b = L/2$, i.e., half of the diffusion column filled with the solute A initially. Then Eq. (4-34) becomes

$$c(l',t_0) = \frac{N_3}{t_0^{1/2}} \left\{ \exp\left[-\frac{(L/4)^2}{Dt_0}\right] - \exp\left[-\frac{(3L/4)^2}{Dt_0}\right] - \exp\left[-\frac{(5L/4)^2}{Dt_0}\right] + \exp\left[-\frac{(7L/4)^2}{Dt_0}\right]\right\}$$

$$= 0.707N_4 \left[\exp\left(\frac{-\pi^2 Dt_0}{4L^2}\right) - \exp\left(\frac{-9\pi^2 Dt_0}{4L^2}\right)\right] \tag{4-35}$$

With $b = L/3$, i.e., two-thirds of the diffusion column filled initially, we have

$$c(l',t_0) = \frac{N_3}{t_0^{1/2}} \left\{ \exp\left[-\frac{(L/6)^2}{Dt_0}\right] - \exp\left[-\frac{(5L/6)^2}{Dt_0}\right] - \exp\left[-\frac{(7L/6)^2}{Dt_0}\right] + \exp\left[-\frac{(11L/6)^2}{Dt_0}\right]\right\}$$

$$= 0.866N_4 \exp\left(-\frac{\pi^2 Dt_0}{4L^2}\right) \tag{4-36}$$

It is interesting to see whether the equations based on the present initial condition, Eq. (4-28), reduce to the equations derived with a pulse initial condition, Eq. (4-5). By letting $b = L - e$, where e is very small, Eq. (4-34) becomes

$$c(l',t_0) = \frac{N_3}{t_0^{1/2}}\left\{\exp\left(-\frac{L}{4Dt_0}\right)\left[\exp\left(\frac{Le/2 - e^2/4}{Dt_0}\right)\right.\right.$$

$$\left. - \exp\left(-\frac{Le/2 + e^2/4}{Dt_0}\right)\right] - \exp\left(-\frac{9L^2}{4Dt_0}\right)$$

$$\left.\cdot\left[\exp\left(\frac{3Le/2 - e^2/4}{Dt_0}\right) - \exp\left(-\frac{3Le/2 + e^2/4}{Dt_0}\right)\right]\right\}$$

$$= N_4\left\{\exp\left(-\frac{\pi^2 Dt_0}{4L^2}\right)\sin\left(\frac{\pi e}{2L}\right) - \exp\left(-\frac{9\pi^2 Dt_0}{4L^2}\right)\right.$$

$$\left.\cdot \sin\left(\frac{3\pi e}{2L}\right)\right\} \qquad (4\text{-}37)$$

Since e is very small, the exponentials within the brackets [ ] can be approximated by the first two terms in a McLaurin expansion, i.e., $\exp(\pm y) \simeq 1 \pm y$, whereas the $\sin(\pi e/2L)$ and $\sin(3\pi e/2L)$ can be set approximately equal to the arguments $\pi e/2L$ and $3\pi e/2L$, respectively. With these approximations and after some rearrangements, the right-hand side after the first equality sign of Eq. (4-37) reduces to the right-hand side of Eq. (4-18), whereas the expression after the second equality sign becomes identical with Eq. (4-26). Both latter equations were derived with a pulse initial condition. In these identifications, c_0ea was put equal to m, the total amount of the solute injected.

The application of Eqs. (4-35), (4-36), or other equivalent to find D from the experimental data is based as before on Eqs. (3-24), (3-25), or (3-27) of Chapter 3, relating $c(l',t_0)$ to the height, h, of the sample peaks. For instance, according to Eq. (3-27), h is simply equal to $2c(l',t_0)$, as given by Eq. (4-35), and the first expression on the right-hand side of Eq. (4-35) predicts that a plot of $\ln(ht_0^{1/2})$ versus $1/t_0$ will be linear at small times, because the exponents in the second, third, and fourth terms are nine or more times bigger than the first, thus making these exponentials negligibly small at small times. However, the determination of D from the slope of such a plot will not be accurate because it requires that b is exactly equal to L/2, so that the slope equals $-(L/4)^2/D$. The same is true for the first expression on the right of Eq. (4-36).

The second expressions on the right-hand side of Eqs. (4-35) and (4-36) are simple exponential functions of t_0 with a well-defined coefficient of t_0, equal to $-\pi^2 D/4L^2$ for the first and nine times bigger for the second exponential. Thus a plot of ln h versus t_0 will be linear for long enough times, so that the second exponential in Eq. (4-35) becomes negligible. Then D can be accurately calculated from the slope.

3. Solution with a Uniform Initial Concentration in Column L

The relevant equations with this initial condition are directly obtained form Eq. (4-34) by putting b = 0:

$$c(l',t_0) = \frac{N_3}{t_0^{1/2}}\left\{1 - 2\exp\left(-\frac{L^2}{Dt_0}\right) + \exp\left(-\frac{4L^2}{Dt_0}\right)\right\}$$

$$= N_4\left\{\exp\left(-\frac{\pi^2 Dt_0}{4L^2}\right) + \exp\left(-\frac{9\pi^2 Dt_0}{4L^2}\right)\right\} \qquad (4\text{-}38)$$

As with Eqs. (4-35) and (4-36), the best use of this equation is to retain only the first term exp $(-\pi^2 Dt_0/4L^2)$ on the far right expression, since the second term becomes negligible compared to that after a short time period. Then, using the relation $h = 2c(l',t_0)$, according to Eq. (3-27), and taking logarithms, we obtain

$$\ln h = \ln (2N_4) - \left(\frac{\pi^2 D}{4L^2}\right)t_0 \qquad (4\text{-}39)$$

It is noteworthy that this same equation is obtained from Eqs. (4-35), (4-36), and (4-27), the only difference being in their preexponential factors. This indicates that the mode of solute introduction into the diffusion column L is irrevelant and does not influence the final equation used to analyze the experimental data. The above conclusion can be checked by introducing varying volumes of the solute and measuring the slope of the ln h versus time plots, neglecting the first few points due to the rest exponential functions.

C. Some Experimental Results

Examples of sample peaks obtained experimentally with gas diffusion as a rate process taking place within the sampling cell, are shown in Figs. 4.3 and 4.4. These should be compared with Figs. 3.5 and 3.6, respectively. The comparison shows that the elution curves predicted by Eqs. (3-21) and (3-19) are confirmed experimentally, apart from the fact that the peaks actually recorded are not square, owing to nonideality.

The height of peaks like those of Fig. 4.3 (measured from the ending baseline) as a function of time is used to plot Eq. (4-22), as was previously mentioned. An example is shown in Fig. 4.5. When the two lengths, l and l', are not equal, two separate plots are required: one with the R- and another with the F-peaks. In Fig. 4.6 Eq. (4-23) was used to plot the height of peaks like those of Fig. 4.4A. Also, examples of plotting data according to Eq. (4-39) are given in Fig. 4.7.

In all cases above, the diffusion coefficient is computed from the slope of the straight lines and the known values of L. Table 4.1 lists some values of D found at ambient temperature, whereas Table 4.2 shows the variation of D with temperature.

The results in Table 4.1 are reduced to 1 atm after multiplication by the pressure, p, of the experiment, given in the table. The actual values of the diffusion coefficient determined experimentally are therefore given by D/p. From the values reported

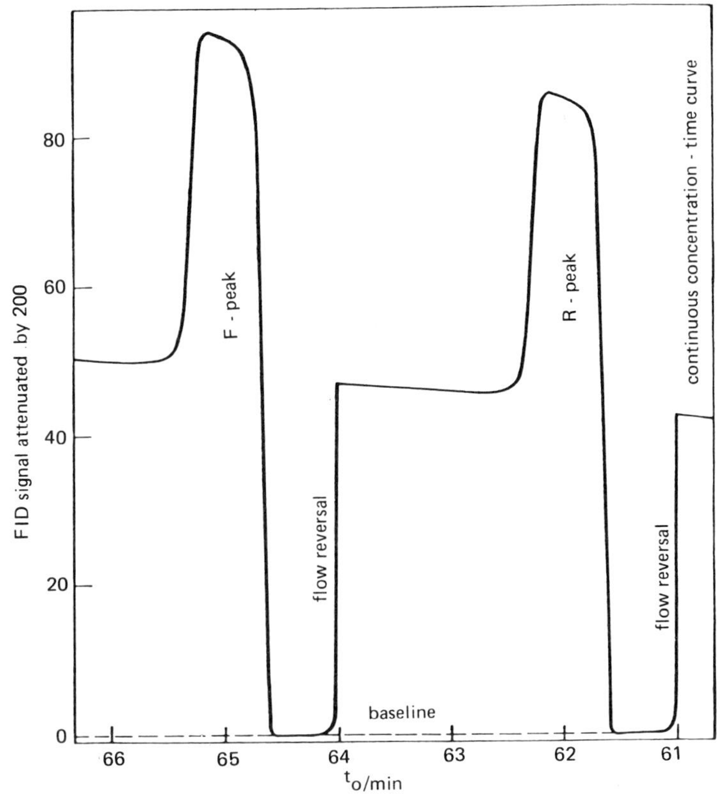

Fig. 4.3 Sample peaks obtained with $t' > t_M + t'_M$, for the diffusion of C_2H_6(0.5 cm^3 pulse) into N_2 ($\dot{V}$ = 0.27 cm^3 s^{-1}), at 293°K and 1.99 atm (1). A straight diffusion column L (61 cm × 4 mm i.d.), and a sampling column l' + l = 40 + 40 cm × 4 mm i.d. were used. (From Ref. 1, used with permission.)

for C_2H_4/N_2 at three different pressures, it is seen that the variation of the results with small changes in pressure (and in $\dot{V}$) is small.

The precision of the present method, defined either as the relative standard deviation (%) or as the relative standard error (%) associated with each value, is as follows. From the values quoted for CH_4/He in Table 4.1, a relative standard deviation

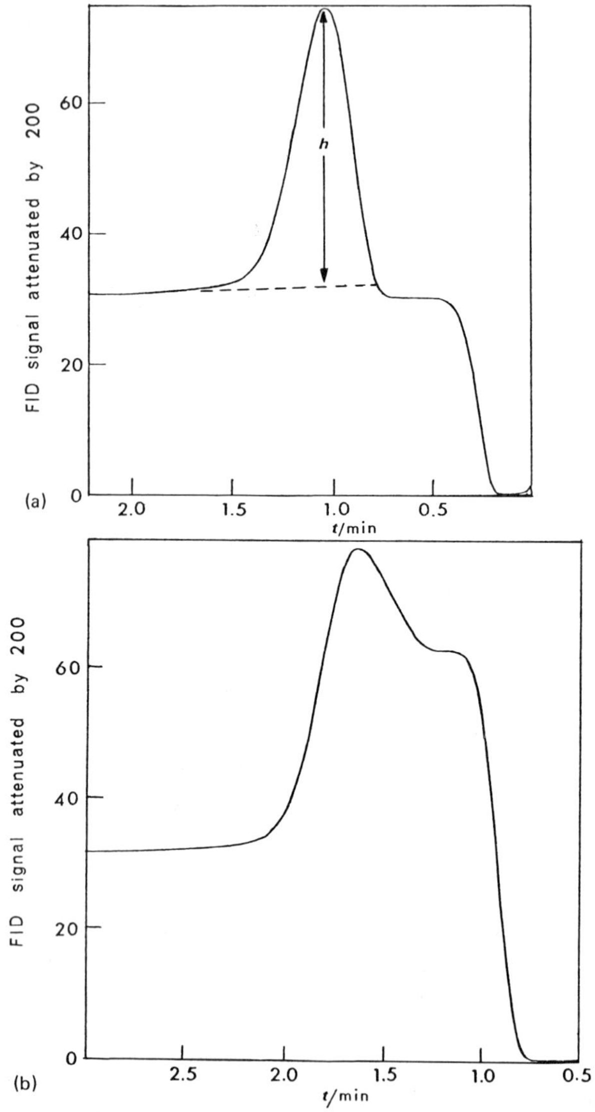

Fig. 4.4 Sample peaks obtained with $t' < t_M + t'_M$, for the diffusion of C_2H_4 (0.5 cm^3 pulse) into He at 373.9 K and $\dot{V}$ = 0.34 $cm^3\ s^{-1}$. The diffusion column L (111.4 cm × 4 mm i.d.) was coiled and placed inside the chromatographic oven. The two lengths, l' + l, of the empty sampling column (99.4 and 99.7 cm × 4 mm i.d.) were also inside the oven. (a) $t' < t_M$ and $t' < t'_M$; (b) $t' > t_M$ and $t' > t'_M$. (From Ref. 2, used with permission.

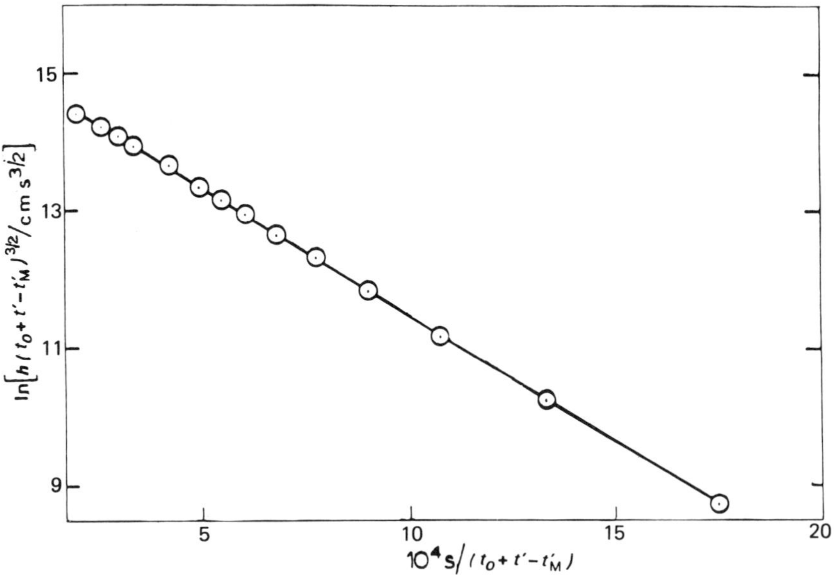

Fig. 4.5 An example of plotting Eq. (4-22) for the diffusion of CH_4 (0.5 cm pulse) into He ($\dot{V}$ = 0.28 $cm^3\ s^{-1}$), at 296°K and 2.03 atm. A straight diffusion column L (61 cm × 4 mm i.d.), and a column l' + l = 40 + 40 cm × 4 mm i.d. were used (1).

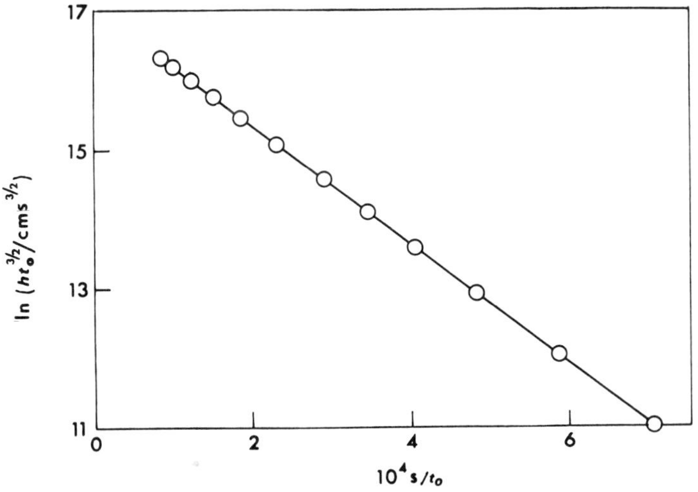

Fig. 4.6 Plot of Eq. (4-23) for the diffusion of C_2H_6 (0.5 cm^3 pulse) into N_2 at 388.5°K and 1 atm, with a coiled diffusion column L (111.4 cm × 4 mm i.d.) placed inside the chromatographic oven. The two lengths, l' + l, were 99.4 + 99.7 cm × 4 mm i.d. (2).

Table 4.1 Diffusion Coefficients of Five Solutes into Three Carrier Gases, at Ambient Temperatures and Reduced to 1 atm Pressure[a]

Solute gas	T (°K)	$\dot{V}$ ($cm^3\ s^{-1}$)	p (atm)	10^3 D ($cm^2\ s^{-1}$) Exp[b]	Calcd[c]	Lit[d]	Accuracy[e] (%)
Carrier gas N_2							
CH_4	296.0	0.260	1.96	272 ± 4	214		21.3
C_2H_6	293.0	0.267	1.99	142 ± 0.03	144	148	1.4(2.7)
n-C_4H_{10}	295.5	0.300	2.15	98 ± 0.2	98.6[f]	96	0.3(2.7)
C_2H_4	296.0	0.120	1.49	168 ± 2	156	163	7.1(4.3)
	292.0	0.268	2.00	156 ± 0.4			0
	292.0	0.538	2.71	161 ± 0.4			3.1
C_3H_6	298.0	0.260	1.96	124 ± 0.4	120[f]		3.2
Carrier gas H_2							
CH_4	293.0	0.287	1.70	699 ± 3	705	730	0.9(3.4)
C_2H_6	297.0	0.267	1.56	548 ± 5	556	540	1.5(3)
n-C_4H_{10}	296.0	0.273	1.60	386 ± 3	373	400	3.4(6.8)

C_2H_4	293.0	0.300	1.75	525 ± 5	559	602	6.5(7.1)
C_3H_6	296.0	0.273	1.60	485 ± 3	486		0.2
Carrier gas He							
CH_4	295.7	0.250	1.78	527 ± 3	669		26.9
	295.0	0.283	2.03	520 ± 1			28.7
	296.0	0.283	2.03	514 ± 0.2			30.2
	296.7	0.283	2.03	522 ± 3			28.2
C_2H_6	295.6	0.300	2.15	518 ± 3	507		2.1
n-C_4H_{10}	290.0	0.283	2.03	333 ± 3	330	364	0.9(9.3)
C_2H_4	296.0	0.283	2.03	558 ± 4	544		2.5
C_3H_6	291.0	0.283	2.03	412 ± 4	440		6.8

[a]Determined using Eq. (4-22). Errors given in Tables 4.1 through 4.9 are standard errors, calculated by regression analysis.
[b]Experimental.
[c]Calculated using the Hirschfelder-Bird-Spotz equation (4a).
[d]Obtained from the literature, i.e., Ref. (6).
[e]Defined by Eq. (4-40). Numbers in parenthesis are the accuracies of the respective literature values.
[f]The necessary parameters σ and ε/k were obtained from Ref. (9c).
Source: Ref. (1).

Table 4.2 Diffusion Coefficients of Three Solutes in Two Carrier Gases at Various Temperatures and 1 atm Pressure[a]

Solute gas	T (°K)	10^3 D (cm^2 s^{-1}) Exp[b]	10^3 D (cm^2 s^{-1}) Calcd[c]	Accuracy[d] (%)
Carrier gas He				
C_2H_6	296.7	491 ± 2	456	7.1
	322.6	556 ± 2	528	5.0
	344.0	618 ± 3	590	4.5
	364.4	684 ± 3	653	4.5
	385.3	745 ± 6	720	3.4
	407.3	807 ± 4	793	1.7
	426.3	878 ± 8	859	2.2
	447.3	941 ± 5	935	0.6
C_2H_4	296.8	525 ± 4	478	9.0
	322.9	599 ± 1	554	7.5
	336.0	649 ± 1	594	8.5
	348.1	674 ± 2	632	6.2
	361.3	726 ± 2	674	7.2
	373.9	780 ± 6	716	8.2
	399.9	860 ± 19	806	6.3
	426.9	932 ± 3	903	3.1
	476.5	1,112 ± 10	1,096	1.4
C_3H_6	345.0	528 ± 0.7	500	5.3
	365.5	584 ± 1	553	5.3
	388.0	642 ± 1	614	4.4
	407.7	690 ± 1	670	2.9
	428.0	750 ± 2	730	2.7
	449.4	819 ± 3	795	2.9
Carrier gas N_2				
C_2H_6	322.8	172 ± 0.2	170	1.2
	345.7	193 ± 0.2	191	1.0
	365.0	214 ± 0.7	210	1.9
	388.5	242 ± 0.3	234	3.3
	407.6	256 ± 0.2	255	0.4
	427.5	282 ± 0.4	277	1.8
	449.3	303 ± 0.5	302	0.3

Table 4.2 (Continued)

Solute gas	T (°K)	10^3 D (cm^2 s^{-1})		Accuracy[d] (%)
		Exp[b]	Calcd[c]	
Carrier gas N_2 (continued)				
C_2H_4	322.8	189 ± 0.08	179	5.3
	344.7	213 ± 0.1	200	6.1
	364.2	234 ± 0.3	221	5.6
	387.6	260 ± 0.3	246	5.4
	407.5	286 ± 0.4	269	5.9
	428.9	306 ± 0.3	294	3.9
	449.8	335 ± 0.9	319	4.8
C_3H_6	322.8	143 ± 0.2	138	3.5
	344.6	164 ± 0.1	155	5.5
	387.4	202 ± 0.2	190	5.9
	406.4	220 ± 0.4	206	6.4
	428.9	243 ± 0.3	227	6.6
	459.0	266 ± 0.2	255	4.1

[a]Determined using Eq. (4-23).
[b]Experimental.
[c]Calculated using the Fuller-Schettler-Giddings equation (5).
[d]Defined by Eq. (4-40).
Source: Ref. (2).

of 1% is calculated. From the standard errors in the same table, a precision better than 1% is found for all but two values (being 1.5 and 1.2%). From the standard errors in Table 4.2, a precision better than 1% is calculated for all but one value (being 2.2%).

Comparison of the diffusion coefficients found with those calculated theoretically, either using the Hirschfelder-Bird-Spotz equation (4a) or the Fuller-Schettler-Giddings equation (5), permits the calculation of the method's accuracy defined as

$$\text{Accuracy (\%)} = \frac{|D_{found} - D_{calcd}|}{D_{found}} \cdot 100 \qquad (4\text{-}40)$$

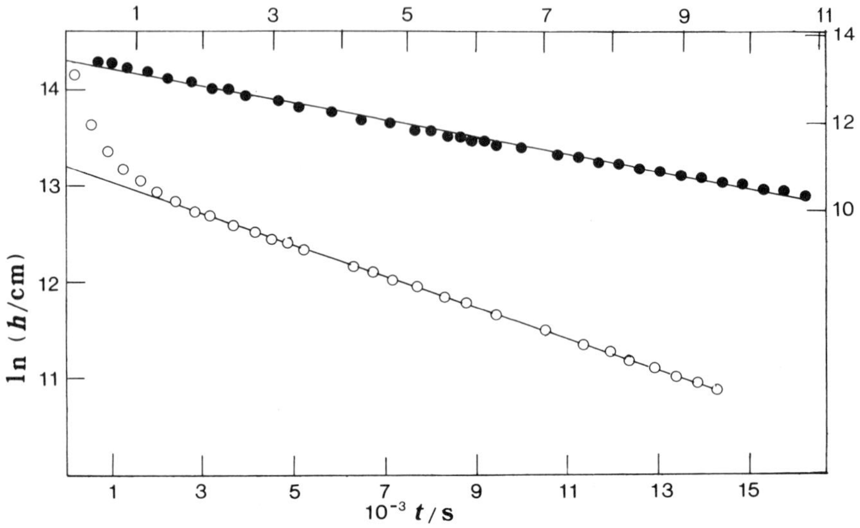

Fig. 4.7 Examples of plotting Eq. (4-39) for the diffusion of 1-butene into He (V = 0.33 cm^3 s^{-1}), at 322.7°K and 1.007 atm. A bent diffusion column incorporated inside the oven and a column l' + l = 40 + 40 cm × 4 mm i.d. were used: o (left ordinate, lower abscissa), points obtained with initial injection of 10 cm^3 1-butene at atmospheric pressure into a column L = 40 cm × 4 mm i.d.; • (right ordinate, upper abscissa), points obtained with 5 cm^2 1-butene at atmospheric pressure injected initially into a column L = 59 cm × 4 mm i.d. (unpublished results).

This is given in the last column of Tables 4.1 and 4.2. With the exception of two pairs containing methane as solute in Table 4.1 and three values for the pair C_2H_4/He in Table 4.2, the accuracy is better than 7.5% in all other 55 cases. The high deviation of the experimental from the calculated values in the methane-containing pairs is probably due to the approximations used in the calculated values. Also, it must be noted that the Fuller-Schettler-Giddings equation (5) gives values closer to those found experimentally than other theoretical equations. Finally, the accuracies of the present method can be compared with those of the values determined by broadening techniques (6), These are given in parenthesis in Table 4.1, and are defined again by

Eq. (4-40) with D_{lit} in place of D_{found}. This comparison leads to the conclusion that with the exception of C_2H_4/N_2 the values of diffusion coefficients determined by the reversed-flow method are closer to the theoretically calculated values than are the experimental values found by broadening techniques under similar conditions of temperature and pressure.

By plotting ln D versus ln T, as shown in Fig. 4.8, we calculated the exponent n in the relationship

$$D = AT^n \tag{4-41}$$

to which all theoretical and semiempirical equations lead for the dependence of D on T. For carrier gases helium and nitrogen, a mean value of 1.61 ± 0.01 and 1.74 ± 0.02, respectively, was found (2). The mean values of n found from similar plots of

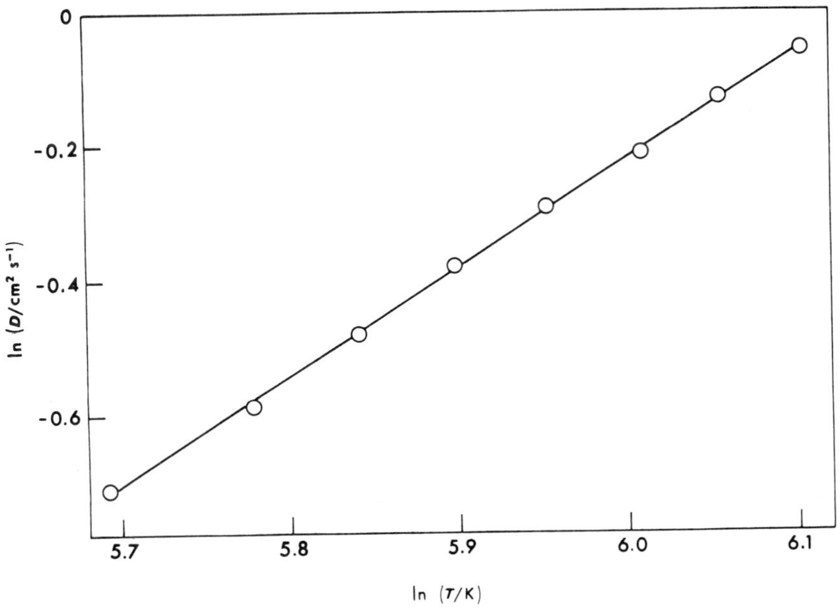

Fig. 4.8 Plot of Eq. (4-41) for the diffusion of C_2H_6 into He (2).

calculated diffusion coefficients (4a) are 1.679 ± 0.001 and 1.808 ± 0.004 for helium and nitrogen, respectively. The mean values for n, from diffusion coefficients determined by the present method, lie between the 1.5, suggested by the Stefan-Maxwell, Gilliland and Arnold equations (7), and 1.81 predicted by the Chen-Othmer equation (7). A value of 1.75 is also predicted by the Huang (6) and the Fuller-Schettler-Giddings equation (5).

II. DIFFUSION COEFFICIENTS IN MULTICOMPONENT GAS MIXTURES

The reversed-flow method for measuring gas diffusion coefficients can be extended to simultaneous determination of diffusion coefficients in multicomponent gas mixtures (8), an experimental problem which has practical as well as theoretical importance. This extension of the method is done by filling part or all of the sampling column (cf. Figs. 3.1 and 4.2) with a chromatographic material, e.g., silica gel, which can effect the separation of some or all components of the gas mixture. When the chromatographic sampling is then performed, preferably by having $t' < t_R$ and $t' < t'_R$, two or more sample peaks appear in the chromatogram (Fig. 4.9). These correspond to two or more different components of the mixture, provided that the components have sufficiently different retention values in the filled sampling column. For each of the components Eqs. (4-23) and (4-39) hold true, and therefore the maximum height, h, of each peak, measured from the ending baseline, can be plotted in the form $\ln(ht_0{}^{3/2})$ versus $1/t_0$ or ln h versus t_0 to yield from the slope its <u>effective</u> binary diffusion coefficient in the mixture. An example is given in Fig. 4.10. Table 4.3 lists some results found (8), together with theoretically calculated values (4a) for the diffusion of each hydrocarbon in pure carrier gas. A comparison between the experimental and the calculated values shows a

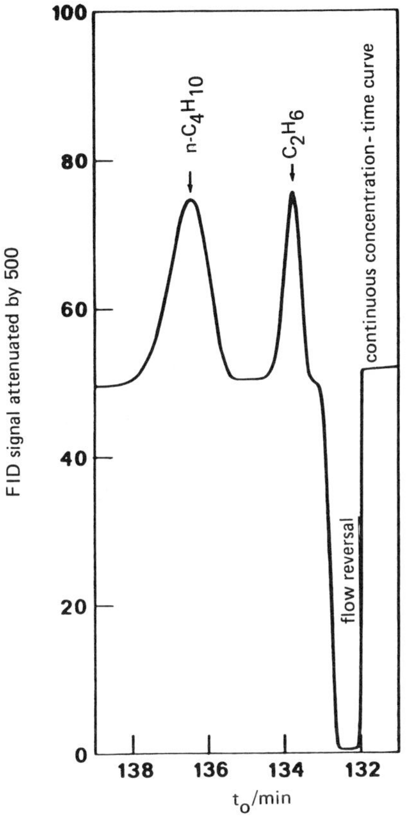

Fig. 4.9 A reversed flow chromatogram for simultaneous determination of diffusion coefficients in a ternary mixture of two solutes (C_2H_6 + n-C_4H_{10}) and a carrier gas (H_2). The temperature of the column L (66.5 cm × 4 mm i.d.) was 292°K and the pressure in it 1.7 atm. The section l (40 cm × 4 mm i.d.) was filled with activated silica gel and held at 454 K. The duration of the backward flow (t') was 24 s (8).

Table 4.3 Effective Diffusion Coefficients Reduced to 1 atm in Some Ternary Mixtures Consisted of a Carrier Gas and Two Hydrocarbons[a]

Carrier gas	Solute gases 1	Solute gases 2	T(°K)	$10^3 D_1$ ($cm^2\ s^{-1}$) Exp[c]	$10^3 D_1$ ($cm^2\ s^{-1}$) Calcd[d]	$10^3 D_2$ ($cm^2\ s^{-1}$) Exp[c]	$10^3 D_2$ ($cm^2\ s^{-1}$) Calcd[d]
H_2	C_2H_4	C_2H_6	298	554 ± 15	593	600 ± 7	557
	C_2H_4	n-C_4H_{10}	296	586 ± 37	584	381 ± 13	379
	C_2H_6	n-C_4H_{10}	292	534 ± 9	538	379 ± 8	372
	C_2H_6	n-C_4H_{10}[b]	292	503 ± 17	538	360 ± 14	372
	C_2H_6		297	556 ± 13	556		
He	C_2H_6	n-C_4H_{10}	294	494 ± 7	506	354 ± 7	338
N_2	C_2H_4	n-C_4H_{10}	296	166 ± 4	156	118 ± 3	99

[a]Experimental values are effective diffusion coefficients in the ternary mixtures, whereas the calculated ones refer to the diffusion in pure carrier gas (8).
[b]The mixture injected was 1.5 cm^3, whereas in all other cases was 0.5 cm^3 (1:1 w/w).
[c]Experimental.
[d]Calculated.

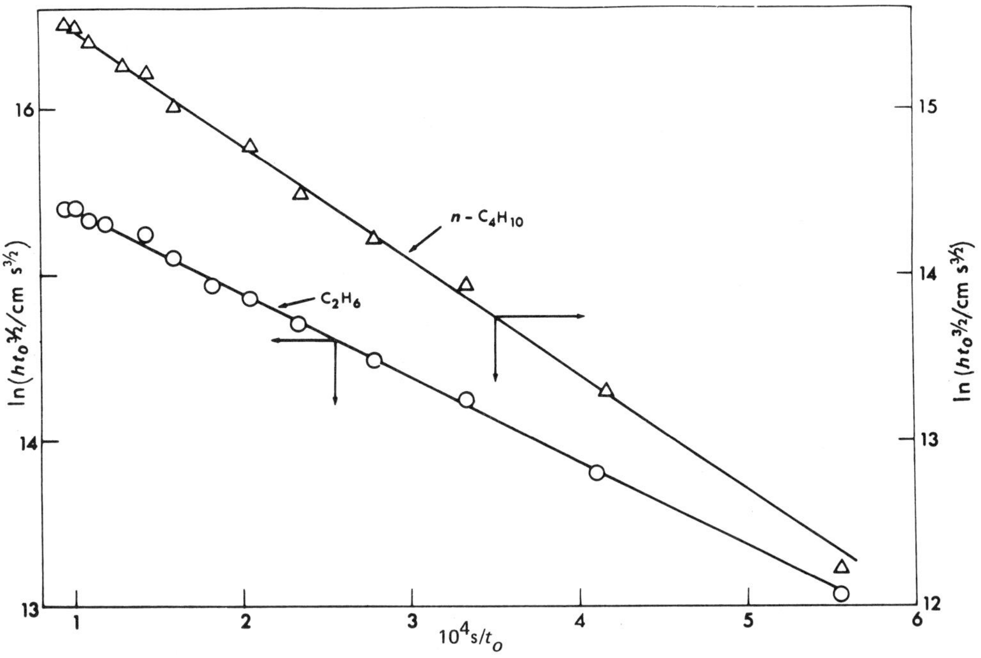

Fig. 4.10 Plot of Eq. (4-23) for the diffusion of C_2H_6 + n-C_4H_{10} into He at 294°K and 2.2 atm. Section 1 of the sampling column (40 cm × 4 mm i.d.) was filled with activated silica gel and held at 454°K. The duration of the backward flow (t') was 24 s. (From Ref. 8, used with permission.)

difference ranging from 0.3 to 7.9%, with one exception (n-C_4H_{10} in N_2), being in that case 16%. These differences are of about the same magnitude as the accuracies for the diffusion of the same hydrocarbons in pure carrier gases (cf. Table 4.1).

These findings are in accord with a limiting case of the Stefan-Maxwell equations (4b) which predict that for small mole fraction of components 1 and 2 in nearly pure carrier gas, the effective diffusion coefficient in the ternary mixture is equal to the diffusion coefficient of each component in pure carrier gas.

The presence of chromatographic material in the sampling column does not seem to influence the results, as shown by a binary mixture (H_2 + C_2H_6) included in Table 4.3. The D_1 found not only coincides with the theoretically calculated value, but also is not significantly different from the value 0.548 cm^2 s^{-1} of Table 4.1, found with the sampling column empty.

III. RELATIVE MOLAR RESPONSES, COLLISION DIAMETERS, AND CRITICAL VOLUMES OF GASES

In another work (10), it has been shown that in the same experiment one can determine together with the diffusion coefficient the relative molar response (RMR) to the thermal conductivity detector. This response can be correlated with fundamental molecular properties of the chromatographic solute and the carrier gas, e.g., their molecular diameters. The collision diameters determined from RMR provide a measure of the molecular diameter of the solute vapor relative to that of the carrier gas. However, these diameters should not be confused with the concept of "true" molecular diameters.

There are several expressions relating the collision diameter of a substance with its critical volume, and hence one can further determine from the collision diameters critical volumes. These are especially important to theoreticians, and are experimentally obtained only by elaborate apparatuses and with great difficulty because no reliable estimation methods have been developed.

The major advantage of the reversed-flow GC method is the estimation of the above four fundamental parameters, i.e., diffusion coefficients, relative molar responses, collision diameters, and critical volumes, for the chromatographic solute under study in one single experiment, with simplicity, short duration, and fairly good accuracy.

A. Theory

According to Eq. (4-23), the plot of ln $(ht_0^{3/2})$ versus $1/t_0$ should give a straight line with slope $b_A = -L^2/4D$ (from which the D values can be found). The intercept of this plot is

$$a_A = \ln (2N_1) = \ln \left[\frac{2mL}{\dot{V}(\pi D)^{1/2}} \right] = \ln \left(\frac{4m}{\dot{V}\pi^{1/2}} \sqrt{-b_A} \right) \qquad (4\text{-}42)$$

The relative molar response, RMR = R_A/R_S, of two gases, A and S, to the detector used can be found by comparing their intercept values, as given by Eq. (4-42). This can be done: 1. by performing two experiments with the two pure gases, masses m_A and m_S, respectively, under constant $\dot{V}$ and other conditions, and 2. by injecting into the diffusion column a mixture of the two gases, A and S (S being every time the same standard, e.g., methane), and filling the column l' + l with a chromatographic material (e.g., silica gel) to obtain two sample peaks for every reversal, corresponding to the two gases. In either case, using Eq. (4-42), one obtains

$$\exp (a_A - a_S) = \frac{R_A}{R_S} \cdot \frac{m_A}{m_S} \left(\frac{b_A}{b_S} \right)^{1/2} \qquad (4\text{-}43)$$

and after rearrangement

$$\frac{R_A}{R_S} = \text{RMR} = \exp (a_A - a_S) \cdot \frac{m_S}{m_A} \left(\frac{b_S}{b_A} \right)^{1/2} \qquad (4\text{-}44)$$

In gas chromatography, the response of the thermal conductivity detector is dependent on the physical properties of the solute A, and publications appeared (11–13) dealing with the prediction of relative molar response factor for this detector. One of them (13) is

$$\mathrm{RMR} = \left(\frac{\sigma_A + \sigma_S}{\sigma_B + \sigma_S}\right)^2 \cdot \left(\frac{M_A - M_S}{M_B - M_S}\right)^{1/4} \tag{4-45}$$

where σ and M indicate molecular diameters and molar masses, respectively. The subscripts A, B, and S refer to the chromatographic solute under study, the carrier gas, and the standard gas (methane), respectively.

By using Eq. (4-45), the molecular diameter of the solute gas, σ_A, can be determined from its RMR value and the known from the literature (4c) molecular diameters of helium (σ_B = 2.576 Å) and methane (σ_S = 3.222 Å):

$$\sigma_A = (\sigma_B + \sigma_S)\left(\frac{M_B - M_S}{M_A - M_S}\right)^{1/8} (\mathrm{RMR})^{1/2} - \sigma_S \tag{4-46}$$

Since the molecular diameter, σ, is immediately related to the critical volume, V_c, by several expressions, the V_c of a substance can be determined from its RMR value with a thermal conductivity detector. Three expressions have the form (11)

$$\sigma = mV_c^{\,n} \tag{4-47}$$

differing only in the pair of the parameters (m,n). The three pairs used are (0.785, 1/3), (0.5894, 0.4006), and (0.561, 5/12). Substituting Eq. (4-47) into Eq. (4-46), one finds

$$V_{cA} = \left[(V_{cB}^{\,n} + V_{cS}^{\,n})\, Q^{1/8}(\mathrm{RMR})^{1/2} - V_{cS}^{\,n}\right]^{1/n} \tag{4-48}$$

where

$$Q = \frac{M_B - M_S}{M_A - M_S} \tag{4-49}$$

Thus using the known values of V_{cB} and V_{cS} (4c), the critical volume of A can be computed from the RMR determined by Eq. (4-44).

B. Experimental Details

1. Apparatus

The apparatus is that shown diagrammatically in Fig. 3.1, with a thermal conductivity detector (TC), without a reference injector, restrictor, or additional separation column. A suitable diffusion column L is 80 cm × 4 mm i.d., whereas the two lengths l' and l, of the sampling column can be 30 cm × 4 mm i.d. each. The carrier gas flowrate must be kept constant in all runs at say 10 $cm^3 min^{-1}$.

The flow reversals are made as usual by switching the four- or six-port valve (cf. Fig. 3.1) from one position (solid lines) to the other position (dashed lines) and vice versa.

2. Procedure

The RMR values are found by performing a pair of separate experiments, with one gas each time without using any chromatographic material in column l' + l. First, pure methane as standard gas is studied, followed by the examined substance, thus avoiding any need for chromatographic separation, and hence increasing the detector sensitivity. The only disadvantage of this procedure is that one must keep exactly the same experimental conditions (flowrate, temperature, and pressure) in the two experiments mentioned above.

For a better accuracy, at least three experiments must be performed for each compound, and the data given in the following Tables 4.4 through 4.7 are average values from these.

Methane was chosen as the standard gas and arbitrarily assigned a signal response of 1 "unit" per mol. Gases with a greater thermal conductivity (closer to helium) than methane would have responses less than 1 unit per mol, whereas those with smaller thermal conductivity would have higher responses.

Table 4.4 Diffusion Coefficients of Some Gases into Carrier Gas Helium at Ambient Temperatures and 1 atm Pressure[a]

		10^3 D (cm^2 s^{-1})		
Solute gas	T(°K)	Exp[b]	Calcd[c]	Accuracy(%)
Ethane	295.8	492 ± 3	508	3.3
Ethene	297.8	550 ± 5	549	0.2
Propane	294.7	450 ± 4	411	8.7
Propene	297	441 ± 2	454	2.9
1-Butene	291	353 ± 2	374	5.9
2-Methylpropene	293	383 ± 3		
cis-2-Butene	291.8	356 ± 2		
trans-2-Butene	292	365 ± 3		
Oxygen	292.2	673 ± 4	716	6.4
Nitrogen	298.4	659 ± 12	702	6.5
Air	294	664 ± 7	694	4.5
Argon	296.4	686 ± 3	722	5.2
Carbon dioxide	297.5	574 ± 2	578	0.7

[a]Determined with a thermal conductivity detector.
[b]Experimental.
[c]Calculated.
Source: Ref. (10), courtesy of Preston Publications, a division of Preston Industries, Inc.

The detailed procedure is that described in section I.A.2 of this chapter. While carrier gas B is flowing in column l' + l, 1 cm^3 of the gas A or the standard gas S is injected into the diffusion column L. After a certain time, during which no signal is noted, an asymmetric concentration-time curve of the gas is recorded. Then, the direction of the carrier gas is reversed for a time period, t', shorter than the gas hold-up time in both column sections l' and l, and then it is restored again to its original direction. After a certain dead time, an extra, fairly symmetrical sample peak is recorded, like sample peak B in Fig. 3.3. This procedure is repeated several times, always with the same dura-

Table 4.5 Relative Molar Responses (RMR) to that of Methane (RMR = 1) for Thermal Conductivity Detector and Helium as Carrier Gas

	RMR		
Solute gas	Exp[a]	Calcd[b]	Accuracy (%)
Ethane	1.420	1.450	2.1
Ethene	1.416	1.346	4.9
Propane	1.767	1.925	8.9
Propene	1.681	1.711	1.8
1-Butene	2.064	2.130	3.2
2-Methylpropene	1.912		
cis-2-Butene	2.094		
trans-2-Butene	2.001		
Oxygen	1.080	1.089	0.8
Nitrogen	1.081	1.137	5.1
Air	1.104	0.934	15.4
Argon	1.101	1.154	4.8
Carbon dioxide	1.242	1.425	14.7

[a]Experimental.
[b]Calculated.
Source: Ref. (10), courtesy of Preston Publications, a division of Preston Industries, Inc.

tion of backward flow, t', giving rise to a series of peaks corresponding to various times from the solute injection.

C. Some Typical Results

As was mentioned before, in a single experiment using a thermal conductivity detector, the diffusion coefficient, the RMR, the collision diameter, and the critical volume of a gas A are determined. In Table 4.4 the diffusion coefficients of some gases into the carrier gas helium, at 1 atm pressure are given. They were determined from the slopes of the plots of ln $(ht_0^{3/2})$ versus $1/t_0$, according to Eq. (4-23). The values found are

Table 4.6 Molecular Diameters of Various Gases, σ_A, Determined by Reversed-Flow GC

Solute gas	σ_A (Å) Found	σ_A (Å) Lit[a]	Accuracy (%)	σ_A (Å) Lit[b]	Accuracy (%)
Ethane	4.347	4.418	1.6	4.384	0.9
Ethene	4.404	4.232	3.9	4.066	7.7
Propane	4.740	5.061	6.8	5.240	10.5
Propene	4.607			4.670	1.4
1-Butene	5.077			5.198	2.4
2-Methylpropene	4.790			4.776	0.3
cis-2-Butene	5.133			5.508	7.3
trans-2-Butene	4.960			5.508	11.1
Oxygen	3.407	3.433	0.8		
Nitrogen	3.526	3.681	4.4	3.85	9.2
Air	3.560	3.617	1.6		
Argon	3.280	3.418	4.2	3.41	4.0
Carbon dioxide	3.559	3.996	12.3	4.00	12.4

[a]Obtained from Ref. (4c).
[b]Obtained from Ref. (9c).
Source: Ref. (10), courtesy of Preston Publications, a division of Preston Industries, Inc.

Table 4.7 Critical Volumes of Various Gases, V_{cA}, Calculated (10) by Eq. (4-48) with the Three Different Values of the Parameter n

	V_{cA} (cm^3 mol^{-1})			Literature		Accuracy (%)		
Solute gas	n=1/3	n=0.4006	n=5/12	Value	Ref.	n=1/3	n=0.4006	n=5/12
Ethane	151.1	142.7	138.1	147.9	(14)	2.1	3.6	7.1
Ethene	158.3	146.0	143.2	133.5	(14)	15.7	8.6	6.8
Propane	200.4	174.8	174.2	192.2	(14)	4.1	10.0	10.3
Propene	183	164	160	181	(14)	1.1	10.4	13.1
1-Butene	250	212	205	241	(14)	3.6	13.7	17.6
2-Methylpropene	207	182	177	240	(14)	15.9	31.9	35.6
cis-2-Butene	259	218	211	236	(14)	8.9	8.3	11.9
trans-2-Butene	232	199	193	236	(14)	1.7	18.6	22.3
Oxygen	67.9	72.8	73.8	74.4	(4c)	9.6	2.2	0.8
Nitrogen	76.1	79.9	80.6	90.1	(4c)	18.4	12.8	11.8
Air	78.6	82.0	82.6	86.6	(4c)	10.2	5.6	4.8
Argon	59.8	65.6	66.8	75.2	(4c)	25.8	14.6	12.6
Carbon dioxide	79	82	83	94	(4c)	19.0	14.6	13.3

Source: Ref (11), courtesy of Preston Publications, a division of Preston Industries, Inc.

Table 4.8 Lennard-Jones Potential Parameters for Three Gas Pairs Calculated from Diffusion Coefficients Determined by the Reversed-Flow GS Technique

Gas pair	T_1(°K)	10^3 D_1 (cm^2 s^{-1})	T_2(°K)	10^3 D_2 (cm^2 s^{-1})	$\varepsilon_{12}k^{-1}$ (°K)	σ_{12} (Å)
C_2H_6-N_2	345.7	193	407.6	256	81.8	4.311
	345.7	193	449.3	303	91.2	4.260
	365.0	214	427.5	282	114.0	4.140
Mean values					95.7	4.237
C_2H_4-N_2	322.8	189	387.6	260	100.1	4.042
	322.8	189	449.8	335	92.0	4.086
	344.7	213	407.5	286	119.8	3.940
Mean values					104.0	4.023
C_3H_6-N_2	322.8	143	459.0	266	124.2	4.311
	344.6	164	387.4	202	130.6	4.237
	344.6	164	428.9	243	156.2	4.124
Mean values					137.0	4.224

Source: Ref. (17), courtesy of Preston Publications, a division of Preston Industries, Inc.

Table 4.9 Lennard-Jones Potential Parameters (ε_1/k and σ_1) and Parachor Values (P) for the Pure Gases Ethane, Ethene, and Propene (17).

Gas	$\varepsilon_1 k^{-1}$ (°K) Found	$\varepsilon_1 k^{-1}$ (°K) Lit. (Ref.)	Deviation (%)	σ_1 (Å) Found	σ_1 (Å) Lit. (Ref.)	Deviation (%)	P Found	P Lit. (Ref.)	Deviation (%)
C_2H_6	192.4	230[a] (9b)	19.5	4.663	4.418[a] (9b)	5.3	134.6	112.2 (14)	16.6
		243[b] (9b)	26.3		3.954[b] (9b)	15.2			
		244[a] (9a)	26.8		3.81[a] (9a)	18.3			
		236[a] (9c)	22.7		4.384[a] (9c)	6.0			
C_2H_4	227.2	205[a] (9b)	9.8	4.204	4.232[a] (9b)	0.7	108.3	101.2 (14)	6.6
		199.2[b] (9b)	12.3		4.523[b] (9b)	7.6			
		222[a] (9a)	2.3		3.70[b] (9a)	12.0			
C_3H_6	394.3	339[a] (9a)	14.0	4.634	4.02[a] (9a)	13.3	158.5	140.2 (14)	11.5
					4.32[b] (9a)	6.8			

[a]Force constants from viscosity.
[b]Force constants from second virial coefficients.
Source: Ref. (17), courtesy of Preston Publications, a division of Preston Industries, Inc.

compared in the same table with those calculated theoretically as before (4a). This comparison permits calculation of the method's accuracy, as defined by Eq. (4-40).

The RMR values, determined by Eq. (4-44), are given in Table 4.5, together with the theoretical ones calculated with the help of Eq. (4-45). In this calculation, the collision diameters of the substances, indicated in Eq. (4-45), have been taken from Ref. (4c), except for C_3H_6 and the four isometric butenes for which the diameters have been taken from Ref. (9c). The accuracy given in the last column is the absolute deviation of the experimental values from those calculated theoretically and is computed by using RMR instead of D in Eq. (4-40).

A straight line is obtained when the RMR in a structurally similar homologous series is plotted against molar mass. An example is given in Fig. 4.11 for the alkenes C_2H_4, C_3H_6, and C_4H_8. The RMR for C_4H_8 is a mean value of those corresponding to 1-butene, 2-methylpropene, cis-2-butene, and trans-2-butene. On the basis of this correlation, it is possible to predict, by a simple calculation, the relative molar response of compounds heretofore unknown, and thus determine their collision diameters and critical volumes. The structural features, e.g., degree of branching, must be considered when such an extrapolation or interpolation is used.

The molecular diameters, computed by Eq. (4-46), are collected in Table 4.6 and compared with the respective literature values from two sources. In most cases, these have been computed from viscosity data. The accuracies given in the same table are defined again by Eq. (4-40) with σ_{found} and σ_{lit} in place of D_{found} and D_{calcd}, respectively. The accuracies are expressed by relatively small numbers, showing the applicability of this technique to measure collision diameters.

Two conclusions, in accord with the theoretical predictions, can be drawn from Table 4.6. First, the molecular diameters

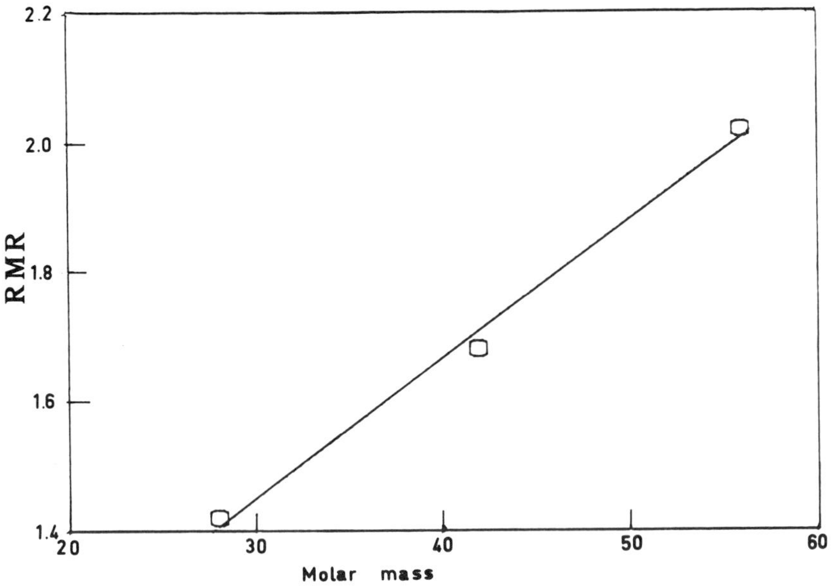

Fig. 4.11 Relative molar response (RMR) to that of methane (RMR = 1) of the thermal conductivity detector for alkenes with 2, 3, and 4 carbon atoms as a function of the molar mass. (From Ref. 10, used with permission of Preston Publications, a division of Preston Industries, Inc.)

increase with the molar mass for compounds belonging to a homologous series. Second, the diameter of a branched compound is smaller than that of the corresponding normal isomer; e.g., 2-methylpropene and 1-butene.

Critical volumes calculated by Eq. (4-48) together with their accuracies defined by an equation analogous to Eq. (4-40), are compiled in Table 4.7. The critical volumes of He and CH_4 were taken from the literature (4c) equal to 57.8 and 99.3 cm^3 mol^{-1}, respectively. By inspection of this table it can be seen that the critical volumes of the inorganic gases are given with a better accuracy by using n = 5/12 in Eq. (4-48). For simple hydrocarbon gases, on the other hand, the value n = 1/3 gives the best accuracy in most cases.

IV. LENNARD-JONES PARAMETERS

The Lennard-Jones (6-12) potential function of intermolecular distance, r_{12}, is given by the relation (9a):

$$\phi(r_{12}) = 4\varepsilon_{12}\left[\left(\frac{\sigma_{12}}{r_{12}}\right)^{12} - \left(\frac{\sigma_{12}}{r_{12}}\right)^{6}\right] \tag{4-50}$$

where $\phi(r_{12})$ is the potential energy of two spherical nonpolar molecules, ε_{12} is the depth of the potential well (the maximum energy of attraction), and σ_{12} is the collision diameter for low-energy collisions, i.e., the value of r_{12} for which $\phi(r_{12}) = 0$. The potential energy of the entire system is found by summing over all pairwise interactions; i.e., $\Sigma\phi(r_{ij})$. The quantities σ_{12} and ε_{12} are the Lennard-Jones potential constants and are related to viscosity, to diffusion, and to second virial coefficients by various relations.

Both parameters σ_{12} and ε_{12} (ε_{12}/k is always used instead of ε_{12}, k = Boltzmann's constant) can be determined from viscosity, from diffusion, and from second virial coefficient measurements. Values of σ_{12} and ε_{12}/k are available from viscosity and from second virial coefficient measurements (4c, 9a-c, 15); values from diffusion data would clearly be preferable, but few are available (9c,15,16). One of the principal reasons for that was the lack of accurate experimental values of diffusion coefficients at various temperatures. Following the determination of these coefficients simply and accurately, as described in the previous sections of this chapter, a treatment of them was undertaken (17) to calculate the Lennard-Jones potential parameters and parachor values. The diffusion coefficient values of the gases ethane, ethene, and propene into carrier gas nitrogen, which were measured at various temperatures by the reversed-flow gas chromatography technique, are those given in Table 4.2. For each binary gas mixture ($C_2H_6-N_2$, $C_2H_4-N_2$ and $C_3H_6-N_2$), three sets of diffusion coefficient values D_{12} (T_1) and D_{12} (T_2) were taken corresponding to three combinations of T_1 and T_2. For every pair of temperatures, the best value of ε_{12}/k was found as follows.

One of the empirical equations, mostly used for the calculation of gas diffusion coefficients, $D_{12}(T)$, at temperature, T, is (4a):

$$D_{12}(T) = 0.0018583T^{3/2} \frac{\left(\frac{1}{M_1} + \frac{1}{M_2}\right)^{1/2}}{p\sigma_{12}^2 \, \Omega_{12}^{(1,1)*}(T^*)} \tag{4-51}$$

where M_1 and M_2 are the molar masses, p is the pressure, $\Omega_{12}^{(1,1)*}(T^*)$ is the first-order collision integral, which takes into account initial relative speeds, reduced mass, and deflection angles, and $T^* = kT/\varepsilon_{12}$.

$\Omega_{12}^{(1,1)*}(T^*)$ is tabulated as a function of kT/ε_{12} for the Lennard-Jones potential in several references (4c,9d), and various analytical approximations are also available, but the author of this work (17) used the accurate relation of Neufeld et al. (18):

$$\omega_{12}^{(1,1)*}(T^*) = \frac{A}{T^{*B}} + \frac{C}{\exp(DT^*)} + \frac{E}{\exp(FT^*)} + \frac{G}{\exp(HT^*)} \tag{4-52}$$

where	A = 1.06036	B = 0.15610
	C = 0.19300	D = 0.47635
	E = 1.03587	F = 1.52996
	G = 1.76474	H = 3.89411

Division of $D_{12}(T_2)$ by $D_{12}(T_1)$ gives

$$\lambda_D = \frac{D_{12}(T_2)}{D_{12}(T_1)} = \left(\frac{T_2}{T_1}\right)^{3/2} \frac{\Omega_{12}^{(1,1)*}(T_1^*)}{\Omega_{12}^{(1,1)*}(T_2^*)} \tag{4-53}$$

Then ε_{12}/k may be determined by trial and error solution of this equation, since λ_D is known from the ratio of the experimental values $D_{12}(T_2)$ and $D_{12}(T_1)$, and $\Omega_{12}^{(1,1)*}(T^*)$ is immediately related to ε_{12}/k via Eq. (4-52). This solution was found by the regula falsi method (19), which is a fast, simple, and accurate iterative method for solving nonlinear equations using a simple computer.

A single $D_{12}(T)$ measurement suffices to determine σ_{12} at a given temperature, T, via Eq. (4-51), once ε_{12}/k is known. The calculated σ_{12} values as well as their mean values with the corresponding values of ε_{12}/k are reproduced in Table 4.8.

As one can see from this table, different sets of force constants were obtained for different choices of the temperatures, T_1 and T_2. This discrepancy, however, is relatively small and results from the fact that the Lennard-Jones potential is an empirical function and does not give an exact description of the dependence of the intermolecular force on the distance, and also from the fact that the various pairs of diffusion coefficients selected have different accuracies.

Using 1. the empirical combining rules

$$\sigma_{12} = \frac{\sigma_1 + \sigma_2}{2} \tag{4-53}$$

and

$$\varepsilon_{12}/k = \left(\frac{\varepsilon_1}{k} \cdot \frac{\varepsilon_2}{k}\right)^{1/2} \tag{4-54}$$

relating force constants between unlike molecules to those between like molecules; 2. the mean values of ε_{12}/k and σ_{12} found; 3. the literature values of ε_2/k and σ_2 for the carrier gas nitrogen, the Lennard-Jones (6-12) force constants for the individual substances were obtained. Values of ε_2/k and σ_2

for nitrogen are available from viscosity, from diffusion, and from second virial coefficient measurements (4c,9a-c). Since ε_{12}/k and σ_{12} were determined from diffusion coefficients, in order to minimize the errors of ε_1/k and σ_1, one must use the values of $\varepsilon_2/k = 47.6$ K and $\sigma_2 = 3.85$ Å, determined from diffusion data (9c).

The Lennard-Jones potential parameters, ε_1/k and σ_1, for the pure gases ethane, ethene, and propene are collected in Table 4.9 and compared with the respective literature values, which have been determined from viscosity and from second virial coefficient measurements. The % deviations given in the same table are defined by a relation similar to Eq. (4-40):

$$\text{Deviation (\%)} = \frac{|X_{\text{found}} - X_{\text{lit}}|}{X_{\text{found}}} \cdot 100 \qquad (4\text{-}55)$$

where X stands for ε_1/k, σ_1, or P (parachor). These deviations in some cases are expressed by relatively big numbers, but in the comparison of ε_1/k and σ_1 values obtained from different data sources, some considerations should be kept in mind.

First, the Lennard-Jones potential is an approximation only of the true intermolecular potential. Second, the use of different properties for the determination of the force constants often results in appreciable differences in the values obtained. Even if the same experimental data are used, differences in two sets of ε_1/k and σ_1 values might result from the use of different averaging procedures.

Such large discrepancies in the values of ε_{12}/k and σ_{12} for gas pairs as well as of ε_1/k and σ_1 for pure gases have also been observed by other workers (15,16).

Since the parachor, P, of a substance is immediately related to the parameters ε and σ by the empirical formula (9e),

$$P = (7.1 \times 10^{23})\varepsilon^{1/4}\sigma^{5/2} \qquad (4\text{-}56)$$

where ε is expressed in erg and σ in cm, one can determine the parachor of the gases used from the values of ε/k and σ, determined as described. Such values with their % deviations, defined again by Eq. (4-55), are given in Table 4.9. The literature values of parachor have been computed from the Sugden formula (14).

As a last conclusion, one can say that the diffusion coefficient values found by the reversed-flow GC method can be used successfully to determine Lennard-Jones potential parameters and parachor values.

LIST OF SYMBOLS

a	Volume of gas phase per unit length of column, or cross-sectional area of void space
a_A, a_S	Intercepts of plot defined by Eq. (4-42) for gases A and S, respectively
b	Length of diffusion column not filled with solute A initially
b_A, b_S	Slopes of plot of $\ln(ht_0^{3/2})$ versus $1/t_0$ for gases A and S, respectively
c	Concentration of a solute vapor in the sampling column
c_z	Concentration of a solute vapor in the diffusion column
c_0	Initial concentration of a solute vapor in the diffusion column
C, C_z	Laplace transforms of c and c_z with respect to t_0
$\bar{C}_z$	Double Lapalce transform of c_z with respect to t_0 and z
D, D_1, D_2, D_{12}	Mutual diffusion coefficients of two gases
e	Length of diffusion column filled initially with solute A

h	Height of a sample peak measured from the ending baseline
l, l'	Lengths of the two sections of the sampling column
L	Length of the diffusion column
m	Amount of a solute injected
M_A, M_B, M_S, M_1, M_2	Molar mass of a gas A, B, S, 1, 2, respectively
N_1, N_2, N_3, N_4	Constants defined by Eqs. (4-17), (4-25), and (4-33), respectively
p_0	Transform parameter with respect to t_0
P	Parachor
q	Parameter defined by Eq. (4-11)
Q	Constant defined by Eq. (4-49)
r_{12}	Intermolecular distance
RMR	Relative molar response defined by Eq. (4-44)
s	Transform parameter with respect to z
t_0	Time from the beginning to the last backward reversal of gas flow
t'	Time interval of backward flow
t_M, t'_M	Gas hold-up time of the empty section l or l', respectively
t_R, t'_R	Ideal retention time on the filled column section l or l', respectively
v	Linear velocity of carrier gas in interparticle space of the sampling column
$\dot{V}$	Volume flowrate of carrier gas
V_c	Critical volume
x, z	Distance coordinates in column $l' + l$ or L, respectively
ε_{12}	Depth of potential well
σ_A, σ_B, σ_S, σ_1, σ_2	Molecular diameter of gas A, B, S, 1, 2, respectively

σ_{12}	Collision diameter
τ	Time defined by Eq. (4-21)
$\phi(r_{12})$	Lennard-Jones potential function, defined by Eq. (4-50)
$\Omega_{12}^{(1,1)*}(T^*)$	First-order collision integral

REFERENCES

1. N. A. Katsanos and G. Karaiskakis, J. Chromatogr., 237: 1 (1982).
2. N. A. Katsanos and G. Karaiskakis, J. Chromatogr., 254: 15 (1983).
3. F. Obberhettinger and L. Badii, Tables of Laplace Transforms, Springer Verlag, New York, 1973, pp. 294, 295, and 421.
4. R. B. Bird, W. E. Stewart, and E. N. Lightfoot, Transport phenomena, Wiley, Chichester, 1960:(a) p. 511; (b) p.570; (c) pp. 744-46.
5. E. N. Fuller, P. D. Schettler, and J. C. Giddings, Ind. Eng. Chem., 58:19 (1966).
6. V. R. Maynard and E. Grushka, Adv. Chromatogr., 12:99 (1975).
7. J. C. Giddings, Dynamics of Chromatography, Marcel Dekker, New York, 1965, p.239.
8. G. Karaiskakis. N. A. Katsanos, and A. Niotis, Chromatographia, 17:310 (1983).
9. J. O. Hirschfelder, C. F. Curtiss, and R. B. Bird, Molecular Theory of Gases and Liquids, Wiley, Chichester, 1954: (a) pp. 552-53; (b) 1110-112; (c) pp. 1212-214; (d) pp. 1126-127; (e) p.356.
10. G. Karaiskakis, A. Niotis, and N. A. Katsanos, J. Chromatogr. Sci., 22:554 (1984).
11. E. F. Barry and D. M. Rosie, J. Chromatogr., 59:269 (1971).
12. E. F. Barry and D. M. Rosie, J. Chromatogr., 63:203 (1971).

13. E. F. Barry, R. Trakimas, and D. M. Rosie, J. Chromatogr. 73:226 (1972).

14. R. R. Dreisbach, Physical Properties of Chemical Compounds—II, Advances in Chemistry Series, Amer. Chem. Soc., Washington D.C., 1959, pp. 12, 13, 217-22.

15. C. E. Cloete, T. W. Smuts, and K. De Clerk, J. Chromatogr., 120:29 (1976).

16. E. Grushka and P. Schnipelsky, J. Phys. Chem., 80:1509 (1975).

17. G. Karaiskakis, J. Chromtogr. Sci., 23:360 (1985).

18. R. C. Reid, J. M. Prausnitz and T. K. Sherwood, The Properties of Gases and Liquids, McGraw-Hill, New York, 1977, p. 549.

19. J. Pachner, Handbook of Numerical Analysis Applications, McGraw-Hill, New York, 1984, Ch. 6.2.

5

Reversed-Flow with a Filled Diffusion Column

I. Introduction

II. Mass Transfer Across a Gas-Liquid Boundary

- A. Theoretical Analysis
- B. Some Experimental Results
- C. Experimental Details

III. Mass Transfer Coefficients for the Evaporation of Liquids and Diffusion Coefficients of Vapors

- A. Theoretical Analysis
- B. Experimental Details
- C. Some Experimental Results
- D. Activity Coefficients

IV. Mass Transfer Across Phase Boundaries Near the Junction of the Diffusion and the Sampling Column

- A. Theoretical Analysis
- B. Some Representative Results

V. Mass Transfer Across Phase Boundaries Extended to All the Diffusion Column

- A. Theoretical Analysis
- B. The Kinetics of Action of Sulfur Dioxide on Marble

List of Symbols

References

I. INTRODUCTION

In the previous chapter various physicochemical measurements were described based on a diffusion column empty of any solid or liquid, and containing only interdiffusing gases. The diffusion current of the solute into the carrier gas stream passing through the sampling column (cf. Fig. 3.1) was shown to be dependent only on the value of the mutual diffusion coefficient of the two gases. Suppose now that the diffusion column contains a solid or a liquid filling part or the entire length, L, of this column, and that the solute gas interacts in anyway with the solid or the liquid. This interaction could be physical (e.g., absorption of the gas by the liquid, adsorption of the gas on the solid, or diffusion into the pores of a porous material); or chemical (e.g., reaction of the solute gas with the solid, or with the liquid, or with substances adsorbed on the solid or dissolved in the liquid).

In all cases above, the rate of change with time of the concentration $c(l',t_0)$ of a solute at the junction $x = l'$ of the sampling cell (cf. Fig. 3.1) will be different from that obtained with an empty diffusion column. This rate of change will now depend not only on the diffusion coefficient of the solute into the carrier gas, but also on the rate coefficient and/or equilibrium coefficient of the interaction between the solute gas and the solid or the liquid contained inside the column L. The rate of change of $c(l',t_0)$ is followed by the usual sampling procedure of reversing the carrier gas flow from time-to-time and measuring the height, h, of the sample peaks thus created as a function of the reversal time, t_0. By using suitable mathematical analysis, we can derive various relations permitting us to calculate from the experimental function $h = g(t_0)$ the rate coefficient and/or the equilibrium coefficient for the gas-solid or gas-liquid interaction. These relations will be derived in the following sections of this chapter for various arrangements, within the diffusion column L, of solids or liquids having various properties and compositions.

The readers who are not interested in the mathematical derivations may skip the purely mathematical sections and arrive directly at the final equations, ready for use with the experimental data.

The three basic arrangements of a solid or liquid in the column L are: 1. A small part of the total column length usually having a larger diameter near its closed end is filled with solid or liquid; 2. a small part of the total length near the junction $x = l'$ is filled; 3. the entire column length, L, is filled with solid particles. In all these arrangements the vapor pressure of the solid or the liquid may be or may not be negligible. This is important only when the solid or liquid vapor is recorded by the detector system. For example, water vapor at a small concentration is not important with a flame-ionization detector.

Let us now examine and analyze some representative cases.

II. MASS TRANSFER ACROSS A GAS-LIQUID BOUNDARY

This problem is important in chemical engineering because it is connected with the unit operation of gas absorption.

A. Theoretical Analysis

1. Formulation of the Problem

Suppose that a small volume (0.5–1 cm^3 at atmospheric pressure) of a solute gas or vapor A is introduced through the injection point of an empty diffusion column L (cf. Fig. 4.2), whereas a carrier gas B flows through the sampling column $l' + l$, filling also column L. The mathematical relations describing the height, h, of the sample peaks as a function of the reversal time, t_0, for this initial condition has been derived in Chapter 4, section I.B, and are given by Eqs. 4-22, 4-23, and 4-26. The latter must be combined with Eq. 3-27:

$$h \approx 2c(l', t_0) \qquad (5\text{-}1)$$

if the duration of the time reversal, t', is smaller than the gas hold-up time in columns l and l'. According to Eq. 4-26, a plot of the height, h, vs. t_0 on a semilogarithmic scale should give what can be termed a diffusion band, and an example is given in Fig. 5.1, curve 1. This band is due solely to the longitudinal diffusion of the solute vapor into the carrier gas; i.e. to one of the main broadening factors in gas chromatography, described

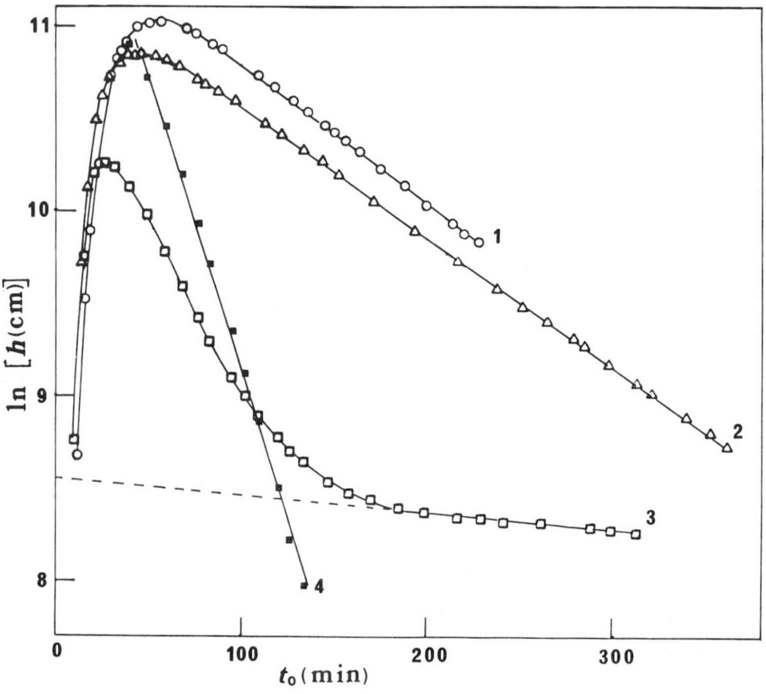

Fig. 5.1 Diffusion bands obtained with 1 cm^3 of butane injected into the diffusion column L at 327°K. Curve 1 was obtained with an L column containing no liquid, curve 2 with L containing 10.4 cm^3 water, and curve 3 with L containing 10.4 cm^3 hexadecane. Line 4 (plotted as 2 + ln h) was obtained by subtracting from the experimental points of curve 3 after the maximum the points corresponding to the extrapolated (dashed line) last linear part of it (Reprinted with permission from Ref. 3. Copyright 1987 American Chemical Society).

in Chapter 1, section IV.C. However, in conventional gas chromatography, the broadening factors are combined with the chromatographic movement of the solute along the column, whereas here the sampling cell used in the reversed-flow method separates the broadening factors from the chromatographic process itself by placing the former in a perpendicular position relative to the latter. Thus gas diffusion is studied alone in the diffusion column which is empty of any solid or liquid material, whereas chromatography is confined to the sampling column and used for separation purposes (when this column is filled) or for the mere creation of chromatographic signals (when the column is empty), either continuous or in the form of extra peaks, by reversing the direction of carrier gas-flow from time-to-time.

Now we come to the second main broadening factor of gas chromatography; i.e., mass transfer phenomena across phase boundaries. These can also be studied separately from the

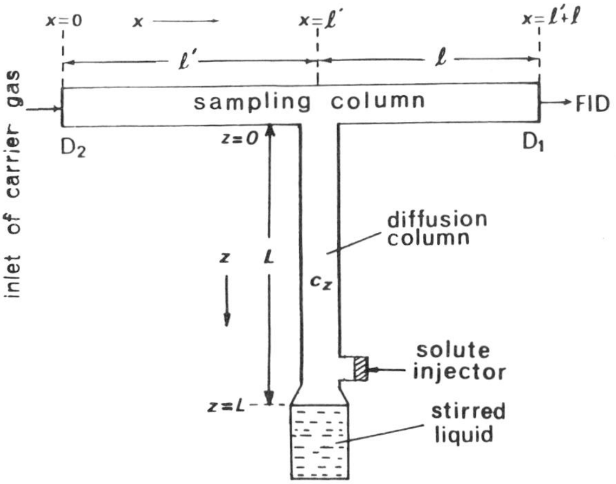

Fig. 5.2 Schematic arrangement of sampling cell in the reversed-flow technique for studying mass transfer phenomena across a gas-liquid boundary at z = L (Reprinted with permission from Ref. 3. Copyright 1987 American Chemical Society).

chromatographic process, inside the diffusion column. Specifically, mass transfer coefficients of a gas across a gas-liquid boundary can be determined by having a small volume of the liquid held at the close end of the diffusion column in a larger diameter section, as shown in Fig. 5.2, the solute gas being introduced as a pulse near the point $z = L$. The rate of change with time of the concentration $c(l', t_0)$ of the solute at the junction $x = l'$ of the sampling cell will be different from that obtained with an empty diffusion column. This rate of change will now depend not only on the diffusion rate of the solute into the carrier gas, but also on its mass transfer rate across the phase boundary created by the presence of the liquid. The rate of change of $c(l', t_0)$ is followed again by the sampling procedure of reversing the carrier gas flow from time-to-time and measuring the height, h, of the sample peaks thus created as a function of time t_0.

The equation describing the diffusion band in the presence of mass transfer across the gas-liquid interphase will now be derived (3), under two assumptions: 1. The vapor pressure of the liquid at the temperature of the experiment is negligible, and 2. diffusion of the solute inside the bulk liquid phase does not intervene. The first assumption is important only when the vapor of the liquid is recorded by the detector system. The second assumption can be substantiated by continuously stirring the liquid. The diffusion equation, Eq. (4-1):

$$\frac{\partial c_z}{\partial t_0} = D \frac{\partial^2 c_z}{\partial z^2} \qquad (5\text{-}2)$$

must be solved under the initial condition

$$c_z(z,0) = \frac{m}{a_G} \delta(z - L) \qquad (5\text{-}3)$$

where m is the amount of solute injected and a_G the cross-sectional area in the columns L and l' + l, subject to the boundary conditions

$$c_z(0,t_0) = c(l',t_0)$$
$$D\left(\frac{\partial c_z}{\partial z}\right)_{z=0} = vc(l',t_0) \qquad (5\text{-}4)$$

at z = 0, and

$$-Da_G\left(\frac{\partial c_z}{\partial z}\right)_{z=L} = K_L a_L(c_L^* - c_L) \qquad (5\text{-}5)$$

at z = L, because the presence of the liquid creates a flux of solute across the gas-liquid boundary, provided the solute dissolves in the liquid. Here a_L is the free surface area of the liquid, c_L the concentration of the absorbed solute in the bulk liquid phase, c_L^* the solute concentration in a fictitious liquid which would be in equilibrium with the real bulk gas phase, and K_L the overall mass transfer coefficient in the liquid phase. This is given by a relation derived in a similar way as that described for analogous relations by Blackadder and Nedderman (1):

$$\frac{1}{K_L} = \frac{1}{k_L} + \frac{K}{k_G} = \frac{K}{K_G} \qquad (5\text{-}6)$$

In this relation, k_G and k_L are the gas and liquid film transfer coefficients, respectively, according to the two-film-theory of Whitman (2); K_G the overall mass transfer coefficient in the gas phase; and K the partition coefficient or distribution constant, giving the relation between equilibrium concentrations in the two phases:

$$K = \frac{c_L^*}{c_z(L,t_0)} \qquad (5\text{-}7)$$

Having in mind that Henry's law constant is $H^{\ddagger} = p/c_L^*$ (p being the partial pressure of the gas) we note that K is analogous and related to $H^{\ddagger}$ by

$$H^{\ddagger} = RT/K \tag{5-8}$$

assuming ideal behavior for the gaseous solute.

To the system of Eqs. (5-2), (5-3), (5-4), and (5-5) one more must be added, describing the rate of change of the absorbed solute concentration, c_L:

$$\frac{\partial c_L}{\partial t_0} = \frac{K_L a_L}{V_L}(c_L^* - c_L) \tag{5-9}$$

where V_L is the volume of the liquid.

2. Solution by Laplace Transformation

Following the same practice as before, the system of partial differential equations (5-2) and (5-9) can be solved by using Laplace transformation of these equations with respect to time t_0 under the initial condition (5-3) and $c_L(0) = 0$ (i.e., no solute absorbed in the liquid initially), together with analogous transformation of the boundary conditions (5-4) and (5-5). As before, we use capital letters C_z, C, C_L, and C_L^* to denote the t_0 – transformed functions of c_z, c, c_L, and c_L^*, respectively, p_0 being the transform parameter. The details of the solution are as follows:

$$p_0 C_z - \frac{m}{a_G}\delta(z - L) = D\frac{d^2 C_z}{dz^2} \tag{5-10}$$

$$p_0 C_L = \frac{K_L a_L}{V_L}(C_L^* - C_L) \tag{5-11}$$

$$C_z(0,p_0) = C(l',p_0) \tag{5-12}$$

$$D\left(\frac{dC_z}{dz}\right)_{z=0} = vC(l',p_0) \tag{5-13}$$

$$-Da_G\left(\frac{dC_z}{dz}\right)_{z=L} = K_L a_L (C_L^* - C_L) \tag{5-14}$$

Eq. (5-10) written in the form

$$\frac{d^2C_z}{dz^2} - q^2C_z = -\frac{m}{a_G D}\,\delta(z - L) \tag{5-15}$$

where

$$q^2 = \frac{p_0}{D} \tag{5-16}$$

can be integrated by using further Laplace transformation with respect to z (transform parameter s):

$$s^2\bar{C}_z - sC_z(0) - C_z'(0) - q^2\bar{C}_z = -\frac{m}{a_G D}\exp(-sL) \tag{5-17}$$

where $\bar{C}_z$ is the double Laplace transform of c_z with respect to t_0 and z, $C_z(0)$ is the t_0 transform of c_z at $z = 0$, and $C_z'(0) = (dC_z/dz)_{z=0}$; i.e., the first z derivative of C_z at $z = 0$. Inverse Laplace transformation of Eq. (5-17) with respect to the parameter s gives

$$C_z = C_z(0)\cosh qz + \frac{C_z'(0)}{q}\sinh qz - \frac{m}{a_G D q}\sinh q(z - L)\cdot u(z - L) \tag{5-18}$$

where $u(z - L)$ is the Heaviside unit step function which equals 0 for $z < L$ and 1 for $z \geq L$.

Now $C_z(0)$ is replaced by $C(l',p_0)$ according to Eq. (5-12), and $C'_z(0)$ by $vC(l',p_0)/D$ according to Eq. (5-13). Further, Eqs. (5-11) and (5-14) are combined to eliminate C_L, and by using also Eqs. (5-6) and (5-7) we obtain

$$-D\left(\frac{dC_z}{dz}\right)_{z=L} = \frac{kp_0}{p_0 + k'}\, C_z(L) \tag{5-19}$$

where

$$k = \frac{K_L K a_L}{a_G} = \frac{K_G a_L}{a_G} \tag{5-20}$$

and

$$k' = \frac{K_L a_L}{V_L} \tag{5-21}$$

Finally, Eq. (5-18) is used to calculate both $(dC_z/d_z)_{z=L}$ and $C_z(L)$ of Eq. (5-19), with the result, after some rearrangement:

$$C(l',p_0) = \frac{m}{a_G D_q}\left[\sinh qL + \frac{v}{Dq}\cosh qL + \frac{kp_0}{Dq(p_0 + k')} \cdot \left(\cosh qL + \frac{v}{Dq}\sinh qL\right)\right]^{-1} \tag{5-22}$$

If there were no liquid in the diffusion column, there would be no mass transfer across a phase boundary. Then, k and k' would be zero and Eq. (5-22) would reduce to Eq. (4-14) of Chapter 4 derived for an empty column L, as expected.

3. Approximations

There remains to find the inverse transformation with respect to p_0 of Eq. (5-22), a more difficult task than the respective problem with Eq. (4-14). It requires again certain approximations, provided of course that the physical conditions permit such approximations to be made. First, we assume as before that for high enough flow rates $v/Dq \gg 1$. Thus, sinh qL can be omitted compared to $(v/Dq)\cosh qL$, and also inside the parentheses ( ) cosh qL is omitted compared to $(v/Dq)\sinh qL$. After that, and some rearrangement, Eq. (5-22) is reduced to

$$C(l',p_0) = \frac{m}{\dot{V}} \operatorname{sech} qL \left[1 + \frac{kp_0}{p_0 + k'} \cdot \frac{\tanh qL}{Dq}\right]^{-1} \tag{5-23}$$

The second approximation is to replace $(m/\dot{V})\operatorname{sech} qL$ and $(\tanh qL)/Dq$ by the following expressions:

$$\frac{m}{\dot{V}} \operatorname{sech} qL \approx \pi m D/\dot{V}L^2(p_0 + \alpha) \tag{5-24}$$

$$(\tanh qL)/Dq \approx 2/L(p_0 + \alpha) \tag{5-25}$$

where

$$\alpha = \frac{\pi^2 D}{4L^2} \tag{5-26}$$

Eq. (5-24) is obtained from (4-26) by retaining only the first exponential function and finding the t_0 Laplace transform of it. In an analogous way, (5-25) is obtained from the first term of the elliptic θ_1 function constituting the image function of $(\tanh qL)/Dq$. Obviously, these approximations are valid at long enough times, so that the second exponential term of the θ_1 function becomes negligible compared to the first term. Using Eqs. (5-24) and (5-25), in Eq. (5-23), one obtains

$$C(l',p_0) = N_2 \frac{p_0 + k'}{(p_0 + k')(p_0 + \alpha) + 2kp_0/L} \tag{5-27}$$

where N_2 is given by Eq. (4-25) as $\pi mD/\dot{V}L^2$.

Taking now the inverse transform of this equation, we find

$$c(l',t_0) = \frac{N_2}{2}\left[\left(1 + \frac{Z_1}{Y_1}\right)\exp\left(-\frac{X_1 + Y_1}{2}t_0\right) + \left(1 - \frac{Z_1}{Y_1}\right)\exp\left(-\frac{X_1 - Y_1}{2}t_0\right)\right] \tag{5-28}$$

where

$$X_1 = \frac{\pi^2 D}{4L^2} + \frac{2K_G a_L}{V_G} + \frac{K_L a_L}{V_L} \tag{5-29}$$

$$Y_1 = \left[\left(\frac{\pi^2 D}{4L^2} + \frac{2K_G a_L}{V_G} + \frac{K_L a_L}{V_L}\right)^2 - \frac{\pi^2 D}{L^2}\cdot\frac{K_L a_L}{V_L}\right]^{1/2} \tag{5-30}$$

$$Z_1 = \frac{\pi^2 D}{4L^2} + \frac{2K_G a_L}{V_G} - \frac{K_L a_L}{V_L} \tag{5-31}$$

and $V_G = a_G L$ is the gaseous volume in column L.

Eq. (5-28) is the mathematical description of a diffusion band destorted by the mass transfer phenomena at long times; i.e., after the time corresponding to the maximum of the band. Comparison of this equation with Eq. (4-26) of a pure diffusion band shows that the exponential function $N_2 \exp(-\alpha t_0)$ has been replaced by the sum of two other functions, one with a bigger and the other with a smaller exponential coefficient than that. Curve 3 of Fig. 5.1 shows an example of such distortion, and this should be compared with the nondistorted curve 1 obtained without the presence of a liquid.

If the two exponential coefficients $(X_1 + Y_1)/2$ and $(X_1 - Y_1)/2$ are sufficiently different, they can be computed from experimental plots like curve 3 of Fig. 5.1; i.e., from plots of ln h vs. t_0. This is done by calculating first the slope $-(X_1 - Y_1)/2$ and the intercept $\ln [N_2(1 - Z_1/Y_1)]$ of the last linear part, and then replotting the initial data of the nonlinear part after the maximum as $\ln \{h - N_2(1 - Z_1/Y_1) \exp [-(X_1 - Y_1)t_0/2]\}$ to find $-(X_1 + Y_1)/2$ from the slope of the straight line obtained, as shown in Fig. 5.1. By <u>multiplying</u> the two exponential coefficients, one finds

$$\left(\frac{X_1 + X_1}{2}\right) \cdot \left(\frac{X_1 - Y_1}{2}\right) = \frac{\pi^2 D}{4L^2} \cdot \frac{K_L a_L}{V_L} \tag{5-32}$$

Dividing this product by the coefficient $\pi^2 D/4L^2$, found without the presence of the liquid with the help of Eq. (4-26), we find the value of $K_L a_L/V_L$, and from the known values of a_L and V_L we calculate K_L, the overall mass transfer coefficient in the liquid phase. The overall mass transfer coefficient in the gas phase K_G is found by <u>adding</u> the two exponential coefficients $(X_1 + Y_1)/2$ and $(X_1 - Y_1)/2$. Their sum equals X_1, and this is given by Eq. (5-29). Then subtracting $\pi^2 D/4L^2$ and $K_L a_L/V_L$ previously found, and using the known values of a_L and V_G, we find K_G. Finally, the partition coefficient, K, is calculated from (5-6) as the ratio K_G/K_L. Thus, from the slopes of the two plots mentioned above, giving the coefficients $(X_1 + Y_1)/2$ and $(X_1 - Y_1)/2$, all coefficients K_L, K_G, and K (and from this also the Henry's law constant $H^{\ddagger}$) can be calculated using simple arithmetic.

From Eqs. (5-29) and (5-30) it is seen that X_1 and Y_1 contain three groups of parameters, $\pi^2 D/4L^2$, $2K_G a_L/V_G$, and $K_L a_L/V_L$. These groups can be termed <u>rate constants</u> for gaseous diffusion, for absorption into the liquid, and for evolution from the liquid, respectively. If $K_L a_L/V_L$ is small compared to the other two, Eq. (5-28) reduces to $c(l',t_0) = N_2 \exp[-(\pi^2D/4L^2 + 2K_Ga_L/V_G)t_0]$; i.e., to a relation resembling Eq. (4-26), but having a bigger exponential coefficient from which K_G can be found. If both $2K_Ga_L/V_G$ and K_La_L/V_L are small compared to $\pi^2D/4L^2$, (4-26) is obtained. This is illustrated by curve 2 of Fig. 5.1.

B. Some Experimental Results

Experiments were carried out with two liquids, hexadecane and water, and three gases, butane, propene, and methane (3). the exponential coefficients (equal to −slopes) of the ln h vs. t_0 plots, obtained from the experimental data, are given in Table 5.1, together with the mass transfer coefficients, partition coefficients, and Henry's law constants calculated from them, as described above. It is seen from Table 5.1 that with hexadecane the two slopes, 1 and 2, for each gas are different by a factor 5-43, and are also sufficiently different from the slope obtained with no liquid in the diffusion column ($V_L = 0$). It has been observed experimentally that better resolution of each curve into two straight lines is achieved with a big volume of liquid. For instance, in the system methane/hexadecane, the two slopes have a ratio 4.6 when $V_L = 10.4$ cm^3, but with 5 cm^3 of liquid, the ratio is much smaller and a plot of only slight curvature is obtained after the maximum (cf. Fig. 5.3). The slope of this plot, considered as a straight line, is $-(1.79 \pm 0.02)10^{-4}s^{-1}$, but no value of K_L, K_G, or K can be calculated from this single slope. To see whether this slope is consistent with those obtained when $V_L = 10.4$ cm^3 was used, the curve with $V_L = 5$ cm^3 was calculated using Eq. (5-28) and the values of X_1, Y_1, and Z_1 computed from the data with 10.4 cm^3 of Table 5.1. The necessary value of N_2 was obtained as the mean of the two relative values of N_2 found from the intercepts of the straight lines with 10.4 cm^3 of liquid, being 6.119×10^4 and 5.808×10^4 (in arbitrary units). The calculated plot, given in Fig. 5.3, is not far from the experimental

Table 5.1 Exponential Coefficients of Eqs. (4-26) and (5-28), Equal to −Slopes of Plots of ln h vs. t_0, Obtained with Two Liquids and Three Gases   Overall Mass Transfer Coefficients in the Liquid (K_L) and in the Gas (K_G) Are Also Given, Together with the Partition Coefficient (K) and Henry's Law[a] Constant ($H^{\ddagger}$)

			10^5(-slope 1)	10^4(−slope 2)				
Gas	T (K)	V_L (cm^3)	s^{-1}	s^{-1}	K_L ($\mu m\ s^{-1}$)	K_G ($\mu m\ s^{-1}$)	K	$H^{\ddagger}$ ($Nm\ mol^{-1}$)
Liquid: hexadecane								
Butane	325.2	0		1.258				
	327.2	10.4	1.630	5.293	2.93	7.65	2.61	1042
	326.2	5	1.827	7.312	2.18	11.6	5.32	509
Propene	324.7	0		1.476				
	325.2	10.4	2.262	7.763	5.08	11.6	2.28	1185
	326.2	5	2.350	9.989	3.27	16.1	4.92	551
Methane	324.7	0		2.375				
	324.7	10.4	9.012	4.141	6.72	2.39	0.355	7595
Liquid: water								
Butane	327.2	10.4		1.164				
	325.2	5		1.314				
Propene	325.2	5		1.510				
Methane	325.7	5		2.401				

[a] The form of the law is $p = H^{\ddagger}c_L^*$.

Source: Ref. (3), with permission, copyright 1987, American Chemical Society.

points. Of course, it is drawn for times after that of the maximum, since Eq. (5-28) was derived and is applied only at such times.

In the other two gases, butane and propene, the values of K_L, K_G, and K obtained with 10.4 and 5 cm^3 hexadecane are of the same order of magnitude and their difference can be attributed to small variations in a_L; i.e., the extent of gas-liquid boundary.

Coming now to the results obtained with water as the liquid phase, it is seen from Table 5.1 that the slope of the single straight line after the maximum differs little from that obtained

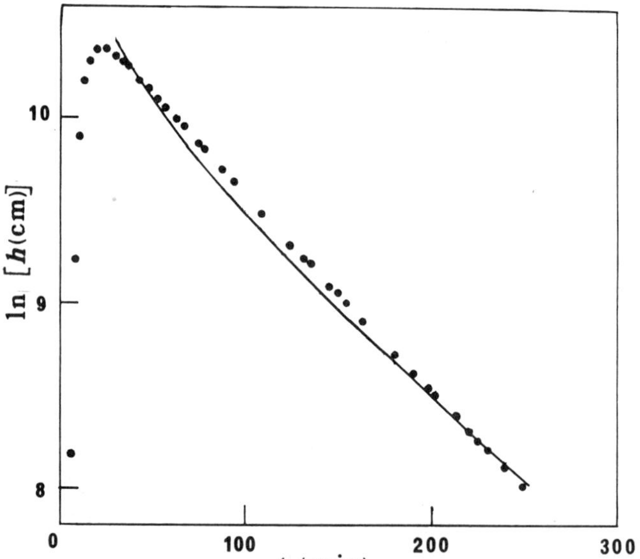

Fig. 5.3 The plot of ln h vs. t_0 for the system methane/hexadecane at 324.7°K with V_L = 5 cm^3. The symbols • are experimental points, while —— is the curve calculated from data obtained with V_L = 10.4 cm^3 (Reprinted with permission from Ref. 3, Copyright 1987 American Chemical Society).

with no liquid in the diffusion column. Therefore, mass transfer phenomena across the boundary hydrocarbon/water are negligible, as expected from the very low solubility of these gases in water at temperatures 52–54°C.

C. Experimental Details

The experimental setup was basically that described elsewhere (4), except for the following. The six-port valve was placed inside a hole in the oven wall. The sections l', l, and L of the sampling cell (cf. Fig. 5.2) were stainless steel 0.25-in. tubes with lengths l' = l = 55 cm and L = 47.5 cm. The liquid was placed in a glass vessel of 17.5 mm i.d. at its lower part and 4 mm i.d. at its upper part, which was connected to the stainless steel column L with a 0.25-in. swagelok union. Three such glass vessels were used: One empty of any liquid, one containing 5cm^3 liquid, and a third containing 10.4 cm^3 liquid. In all three cases, the distance between the axis of the solute injector and the liquid surface or the bottom of the empty vessel was 2 cm. The liquid was stirred throughout the whole experiment by means of a small magnetic stirrer using a glass-coated stirring bar.

The gas volume, V_G, in column L was 10.60–10.92 cm^3, as measured by filling it with water at a certain temperature, weighing it, and using the density of water at that temperature.

The pressure drop along l' + l was negligible, and the pressure inside the whole cell was 1 atm. The carrier gas flowrate (corrected at column temperature) was 0.36 cm^3 s^{-1}.

As a conclusion from section II, one can say that by using a very simple experimental arrangement and a simple mathematical analysis, the reversed-flow gas chromatographic technique leads to the determination of mass transfer coefficients across a gas-liquid boundary, together with the partition coefficient (or the Henry's law constant) for the distribution of a solute between the gas and the liquid phases. Practical applications of these coefficients can be found, e.g., in gas chromatography, gas absorption, and evaporation.

III. MASS TRANSFER COEFFICIENTS FOR THE EVAPORATION OF LIQUIDS AND DIFFUSION COEFFICIENTS OF VAPORS

When studying mass transfer coefficients across a gas-liquid boundary (section II of this chapter) it was assumed that either the vapor pressure of the liquid was negligible or the vapor of the liquid was a nondetectable one by the chromatograph detector system. In the present section, we shall examine the case where the detectable solute in the carrier gas is the vapor coming out from the liquid contained again at the lower part of the diffusion column, as in Fig. 5.2. No foreign vapor or gas, however, is injected into the column. Just the vapor of the liquid (nonagitated) diffuses along the length, L, into the carrier gas, passing again through column l' + l (Fig. 5.4).

A. Theoretical Analysis

Here the diffusion equation (5-2) must be solved under the initial condition

$$c_z(z,0) = 0$$

subject to boundary conditions at z = 0, those described by Eqs. (5-4), and boundary condition at z = L the relation

$$D\left(\frac{\partial c_z}{\partial z}\right)_{z=L} = K_G[c_0 - c_z(L)] \tag{5-33}$$

where D is the diffusion coefficient of the vapor into the carrier gas, K_G the overall mass transfer coefficient for the solute

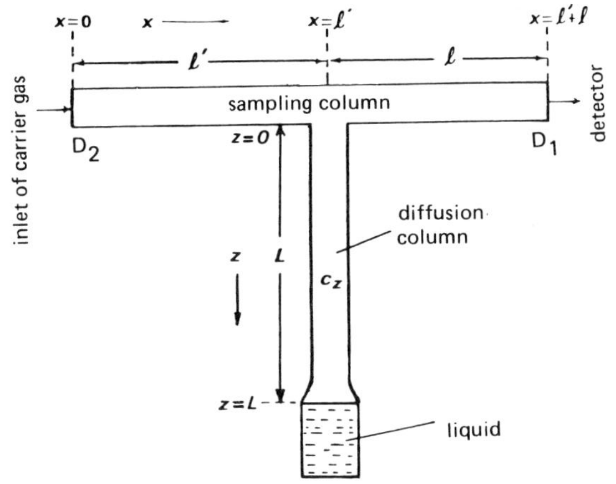

Fig. 5.4 Sampling cell for studying mass transfer coefficients and diffusion coefficients of the vapors coming out from a liquid.

evaporation, $c_z(L)$ the actual vapor concentration at the liquid interphase, and c_0 the concentration of the vapor which would be in equilibrium with the bulk liquid phase. Eq. (5-33) expresses the equality of the diffusion flux for removal of vapors from the liquid surface and the evaporation flux due to departure of c_z at the surface from the equilibrium value, c_0. It is written for a free surface area of liquid, a_L, equal to the cross-sectional area, a_G, of column L. If these two areas are different, the right-hand side of Eq. (5-33) must be multiplied by the ratio a_L/a_G.

The solution of Eq. (5-2) under the above conditions, as regards the independent variable z, is

$$C(l',p_0) = \frac{K_G c_0}{Dqp_0} \left[\sinh qL + \frac{V}{Dq} \cosh qL + \frac{K_G}{Dq}\left(\cosh qL + \frac{V}{Dq} \sinh qL\right)\right]^{-1} \tag{5-34}$$

where $C(l',p_0)$ is the Laplace transformed function $c(l',t_0)$ with respect to t_0, and q is given by Eq. (5-16). This solution is based on the assumption that the composition of the liquid does not change appreciably during a kinetic run, since a very small volume of it evaporates until a constant infinity value in the detector signal is reached. The usual approximation $v/Dq \gg 1$ now, reduces the equation to

$$C(l',p_0) = \frac{K_G c_0}{v p_0} \left(\cosh qL + \frac{K_G}{Dq} \sinh qL \right)^{-1} \tag{5-35}$$

Inverse Laplace transformation of this equation to find $c(l',t_0)$ can be achieved by using two other approximations, one for small and one for long times (4). In the first case, qL is large, allowing both cosh qL and sinh qL to be approximated by exp (qL)/2. Then Eq. (5-35) becomes

$$C(l',p_0) = \frac{2K_G c_0}{vD} \cdot \frac{\exp(-qL)}{q(q + K_G/D)} \tag{5-36}$$

and its inverse transform is

$$c(l',t_0) = \frac{2K_G c_0}{v} \cdot \exp\left(\frac{K_G L}{D} + \frac{K_G^2 t_0}{D} \right) \cdot$$

$$\cdot \operatorname{erfc}\left[\frac{L}{2(Dt_0)^{1/2}} + K_G \left(\frac{t_0}{D} \right)^{1/2} \right] \tag{5-37}$$

Finally, if one uses the relation

$$\operatorname{erfc} \approx \frac{\exp(-x^2)}{x\pi^{1/2}} \tag{5-38}$$

which is a good approximation for large values of x (5), Eq. (5-37) becomes

$$c(l',t_0) = \frac{2K_G c_0}{v}\left(\frac{D}{\pi}\right)^{1/2} \cdot \exp\left(-\frac{L^2}{4Dt_0}\right) \cdot \left(\frac{L}{2t_0^{1/2}} + K_G t_0^{1/2}\right)^{-1} \tag{5-39}$$

Coming now to the other extreme, i.e., long time approximations, qL is small, and the functions cosh qL and sinh qL of Eq. (5-35) can be expanded in a Maclaurin series, retaining only the first three terms in each of them. Then one obtains by using (5-16) and making some rearrangements

$$C(l',p_0) = \frac{2K_G D c_0}{vL^2 p_0} \cdot \frac{1}{p_0 + 2(K_G L + D)/L^2} \tag{5-40}$$

Inverse transformation of this relation yields

$$c(l',t_0) = \frac{K_G D c_0}{v(K_G L + D)}\left\{1 - \exp\left[-\frac{2(K_G L + D)t_0}{L^2}\right]\right\} \tag{5-41}$$

By considering the maximum height, h, of the sample peaks in Eq. (5-1) and substituting in it the right-hand side of Eq. (5-41) for $c(l',t_0)$, one obtains h as an explicit function of time t_0. In order to linearize the resulting relation, an infinite value h_∞ for the peak height is required:

$$h_\infty = \frac{2K_G D c_0}{v(K_G L + D)} \tag{5-42}$$

Using this expression, we obtain

$$\ln (h_\infty - h) = \ln h_\infty - \frac{2(K_G L + D)}{L^2} \cdot t_0 \tag{5-43}$$

Thus, at long enough times, for which Eq. (5-41) was derived, a plot of ln $(h_\infty - h)$ versus t_0 is expected to be linear, and from the slope $-2(K_G L + D)/L^2$ a first approximate value of K_G can be calculated using the known value of L and a literature or theoretically calculated (6) value of D. This value of K_G can be used to plot small times data according to Eq. (5-39), which is substituted now for $c(l', t_0)$ in Eq. (5-1). After rearrangement and by taking logarithms, there results:

$$\ln\left[h\left(\frac{L}{2t_0^{1/2}} + K_G t_0^{1/2}\right)\right] = \ln\left[\frac{4K_G c_0}{\dot{V}}\left(\frac{D}{\pi}\right)^{1/2}\right] - \frac{L^2}{4D} \cdot \frac{1}{t_0} \qquad (5\text{-}44)$$

Now a plot of the left-hand side of this relation versus $1/t_0$ will yield a first approximation experimental value for D from the slope $-L^2/4D$ of this new linear plot. This D value can be used back in the slope found from the plot of Eq. (5-43) to calculate a more accurate value for K_G. In turn, the latter is utilized to replot Eq. (5-44), so that a more accurate value for D is found. These iterations can be continued until no significant change in the K_G and D values results.

B. Experimental Details

The apparatus being used in the present problem has been described (4) and is given in Fig. 5.5.

After placing the liquid under study in its position and waiting for a certain time, during which no signal is noted, one records a monotonously rising concentration-time curve for the vapor of the liquid. When this continuous rising signal is high enough, the chromatographic sampling procedure is started by reversing the direction of the carrier gas flow for a time period

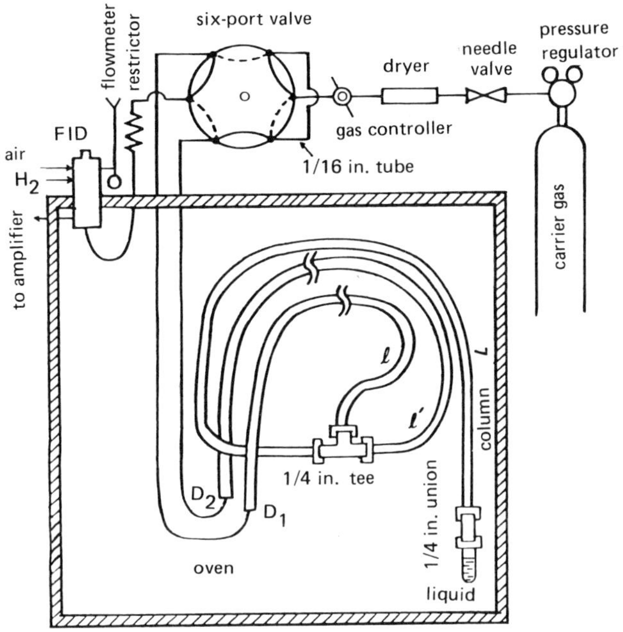

Fig. 5.5 Gas lines and important connections for measuring simultaneously diffusion coefficients and mass transfer coefficients for evaporation (Reprinted with permission from Ref. 4, copyright 1984 American Chemical Society).

shorter than the gas hold-up time in both column sections, l and l'. Then the gas flow is restored to its original direction.

The pressure drop along column l' + l is ususally negligible, and the pressure inside the whole cell can be measured by an open mercury manometer. A carrier gas flowrate in the range 0.25–1 $cm^3\ s^{-1}$ is used.

C. Some Experimental Results

1. Pure Liquids

Fig. 5.6 shows a section of a typical chromatogram with sample peaks created by flow reversals. In Fig. 5.7, the height, h, of the sample peaks a a function of the time t_0, when the flow

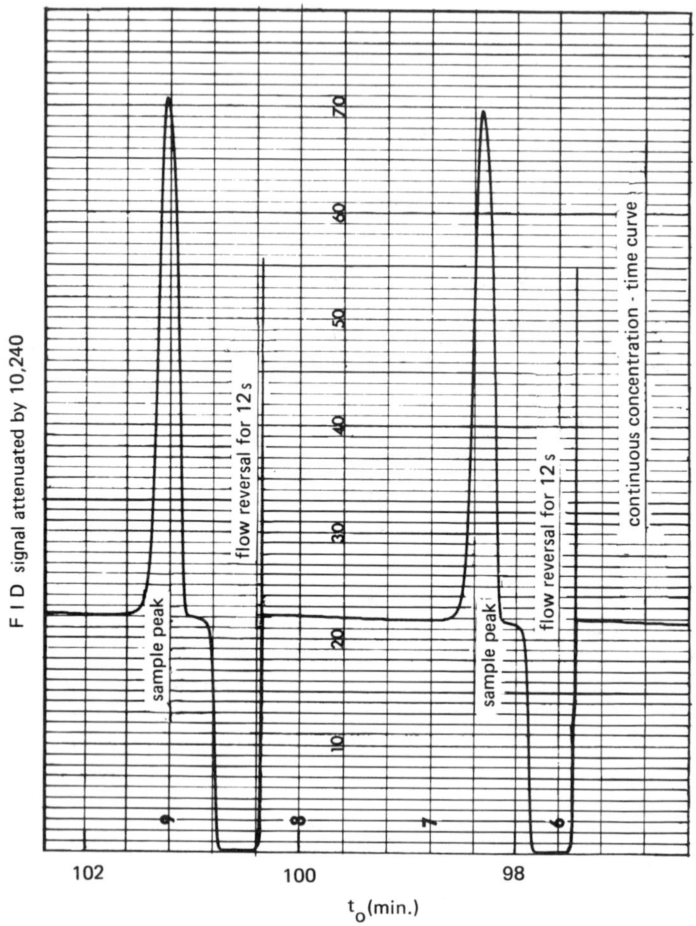

Fig. 5.6 A reversed-flow chromatogram showing two sample peaks for the diffusion of 1-propanol vapor into helium at 342.4°K and 1 atm with $\dot{V}$ = 0.635 cm^3 s^{-1} (12).

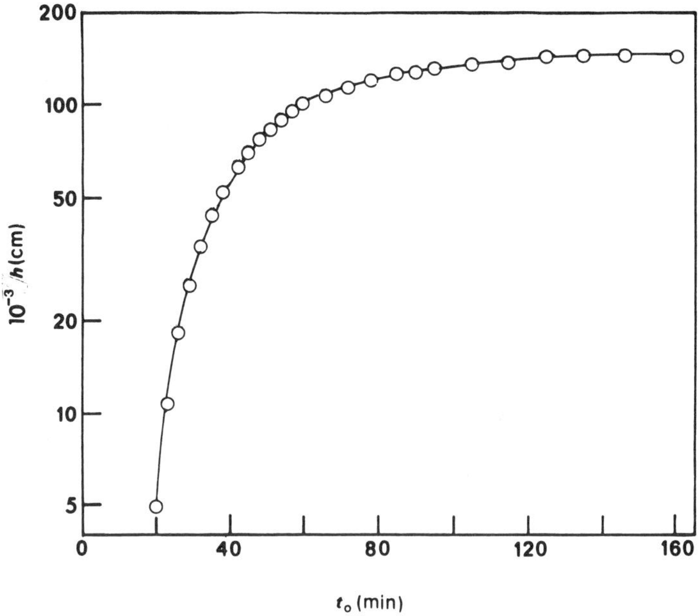

Fig. 5.7 The rise of the sample peak height with time for the diffusion of ethanol vapor into helium ($\dot{V}$ = 0.543 $cm^3\ s^{-1}$), at 336.8 K and 1 atm (Reprinted with permission from Ref. 4, copyright 1984 American Chemical Society).

reversal was made, is plotted on a semilogarithmic scale. It shows the steep rise and then the leveling off with time of the sample peak height.

As an example of using Eqs. (5-43) and (5-44) to analyze the experimental findings, the data of Fig. 5.7 are treated as follows. Leaving out the first 3–4 points, which correspond to small times, the rest of the experimental points are plotted according to Eq. (5-43), as shown in Fig. 5.8. As infinity value h_∞ was taken the mean of the values found in the time interval 135–160 min, which differed little from one another. From the slope of this plot, which is equal to $-2(K_G L + D)/L^2$, according to Eq. (5-43), using a theoretically calculated (6) value of

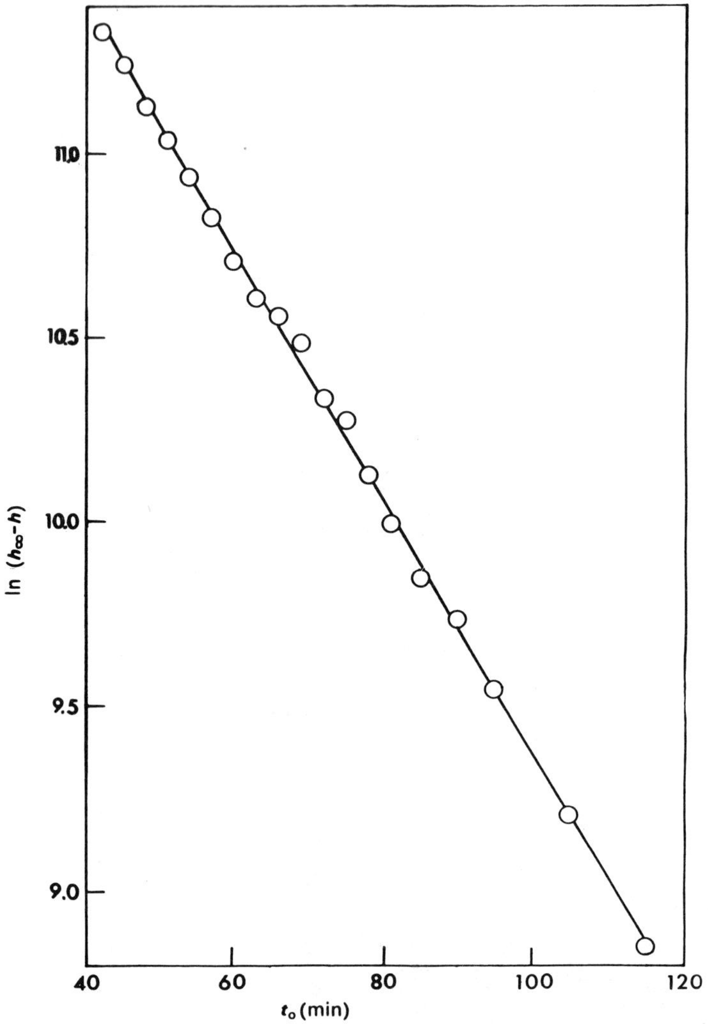

Fig. 5.8 Plot of Eq. (5-43) for the data of Fig. 5.7 (Reprinted with permission from Ref. 4, copyright 1984 American Chemical Society).

$D = 0.593\ cm^2\ s^{-1}$, and the actual value of L (116.2 cm), a value of $278 \times 10^{-4}\ cm\ s^{-1}$ for K_G is calculated. This approximate value is now used to plot all but the few points close to h_∞ according to Eq. (5-44), as shown in Fig. 5.9. From the slope of this latter plot, which equals $-L^2/4D$, as shown in the Theoretical section III.A, a value of $0.576\ cm^2\ s^{-1}$ for D is found. If this is combined with the slope of the previous plot (Fig. 5.8), a second value for $K_G = 279 \times 10^{-4}\ cm\ s^{-1}$ is calculated and further used to replot the data according to Eq. (5-44). The new value for D found coincides with the previous one ($0.576\ cm^2\ s^{-1}$), and thus the iteration procedure must be stopped.

Table 5.2 summarizes the results obtained with some pure liquids (4) together with the diffusion coefficients calculated

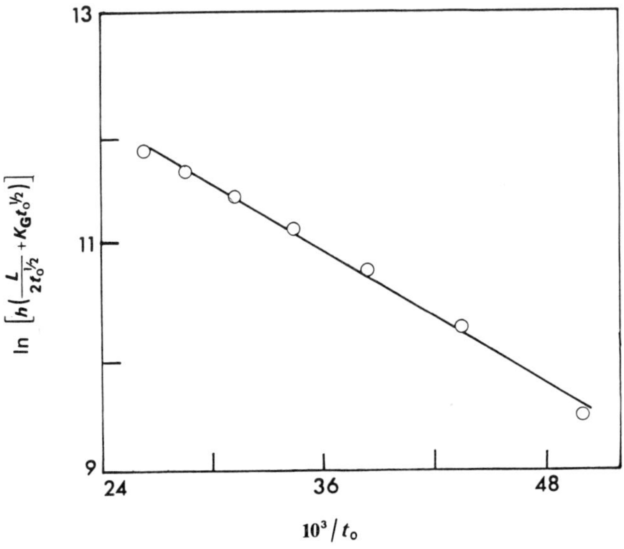

Fig. 5.9 Plot of Eq. (5-44) for the data of Fig. 5.7 (Reprinted with permission from Ref. 4. Copyright 1984 American Chemical Society).

Table 5.2 Mass Transfer Coefficients for Evaporation of Pure Liquids, and Diffusion Coefficients of Vapors into Helium at 1 atm

				10^3 D (cm^2 s^{-1})			
Liquid	$\dot{V}$ (cm^3 s^{-1})	T (K)	$10^4 K_G$ (cm s^{-1})	Ref. (4)	Calcd	Ref. (7)	Accuracy[c] (%)
n-Pentane	0.362	300.7	369	462	340	292	26.4(16.4)
n-Hexane	0.295	295.2	228[a]	334[a]	311	268	6.9(16)
	0.543	293.2	224	300	308	265	2.7(16.2)
	0.532	332.5	279	377	381	331	1.1(15.1)
	0.290	293.2	257[b]	262[b]	308	265	17.6(16.2)
n-Heptane	0.287	342.7	257	332	343		3.3
n-Octane	0.362	322.8	251	254	265	245	4.3(8.2)
	0.532	373.9	340	320	340	317	6.3(7.3)
Methanol	0.287	322.5	330	705	688	642	2.4(7.2)
	0.602	322.5	340	663	688	642	3.8(7.2)
Ethanol	0.543	336.8	279	576	593	614	3(3.4)
Propan-1-ol	0.287	342.3	364	473		467	1.3
Butan-1-ol	0.287	358.1	268	423		438	3.5
Pentan-1-ol	0.287	374.0	203	398		409	2.8

[a]Values determined by using a diffusion column with length L = 86.5 cm.
[b]Values determined by using a diffusion column with length L = 120.2 cm.
[c]Defined by Eq. (5-45). Numbers in parentheses are the accuracies of the respective literature values.

theoretically (6) and those found in the literature (7). In the latter case, the values were reduced to the temperature of Table 5.2 by using the relation $D_1/D_2 = (T_1/T_2)^{1.75}$, as suggested by the Fuller-Schettler-Giddings equation (8).

As one can see from Table 5.2, the D values determined (4) are very close to the theoretical ones or those found in the literature. The accuracy given in the last column is the deviation of the experimental values from those calculated theoretically, except for the last three cases, in which instead of the D_{calcd}, we used the literature values because the appropriate constants for the determination of D_{calcd} were not available:

$$\text{Accuracy (\%)} = \frac{|D_{found} - D_{calcd}|}{D_{found}} \times 100 \tag{5-45}$$

With the exception of two pairs, this accuracy is better than 7% in all cases. The accuracies of the respective literature values are expressed by bigger numbers, except for two cases.

A comparison of the K_G values determined with others in the literature is impossible because such values could not be found. With respect to the values of Table 5.2, two tentative conclusions can be drawn. First, the K_G is independent of the carrier gas flowrate, and second, these coefficients increase with increasing temperature.

The boundary condition (5-33), adopted from Crank (9), is based on a departure from equilibrium at the liquid-gas interphase. This contrasts with the assumption of equilibrium at the interphase of Bird et al. (10), but the latter is not supported by the experimental findings. Thus if Eq. (5-33) is abandoned and the boundary condition $c_z(L) = c_0$, corresponding to equilibrium, is used, we obtain the relation

$$C(l',p_0) = \frac{c_0}{p_0} \cdot \frac{1}{(v/Dq)\sinh qL + \cosh qL} \tag{5-34'}$$

instead of Eq. (5-34). Adopting the same approximations as before, we find for small times

$$c(l',t_0) = \frac{2c_0}{v}\left(\frac{D}{\pi}\right)^{1/2} \cdot \exp\left(-\frac{L^2}{4Dt_0}\right) \Big/ t_0^{1/2} \tag{5-39'}$$

an equation which could also be obtained directly from Eq. (5-39) by assuming a large value for K_G. Plots of the experimental data as ln $(ht_0^{1/2})$ versus $1/t_0$, according to Eq. (5-39'), show that this equation does not fit the data so well as does Eq. (5-44) derived from Eq. (5-39). This is evidenced by the poorer linearity of the plots and the lower accuracy of diffusion coefficients calculated from their slopes.

Coming to the long time approximation of Eq. (5-34'), as we did for Eq. (5-34), we derive the relation

$$c(l',t_0) = \frac{Dc_0}{vL + D}\left\{1 - \exp\left[-2\left(\frac{D}{L^2} + \frac{v}{L}\right)t_0\right]\right\} \tag{5-41'}$$

in place of Eq. (5-41). Instead of Eq. (5-43), we would then have

$$\ln (h_\infty - h) = \ln h_\infty - 2\left(\frac{D}{L^2} + \frac{v}{L}\right)t_0 \tag{5-43'}$$

and the plots of $\ln (h_\infty - h)$ versus t_0, as that of Fig. 5.8, would have a slope $-2(D/L^2 + v/L)$ which can be predicted from the known values of D, L, and v. For example, the slope of the plot in Fig. 5.8 would be $-0.074\ s^{-1}$, whereas that found from the experimental plot is $-5.65 \times 10^{-4}\ s^{-1}$; i.e., two orders of magnitude smaller. The same applies to the slope of the plots for all other substances studied. Moreover, Eq. 5-43 predicts independence of the slope from carrier gas flowrate, whereas Eq. (5-43') predicts a linear dependence of the slope on the flowrate. The experimentally determined slopes are independent of the flowrate, and thus in agreement with Eq. (5-43).

In conclusion, the assumption of equilibrium at the liquid-gas interphase leads to profound disagreement with the experimental evidence.

2. Liquid Mixtures

The mass transfer coefficients for the evaporation of the liquid mixtures methanol/water, ethanol/water, and propan-1-ol/water as well as the diffusion coefficients of the alcohol's vapors into the carrier gas helium have also been determined (11).

Table 5.3 summarizes the results obtained with all liquid mixtures studied at one temperature and various alcohol mole fractions. Comparison of the diffusion coefficients found with those calculated theoretically, using the Hirschfelder-Bird-Spotz equation (6), permits the calculation of the method's accuracy defined by Eq. (5-45). This is given in the last column of Table 5.3.

As one can see from Table 5.3, the K_G values increase with increasing alcohol mole fraction X_A. In all cases there is a steep rise in K_G and then a leveling off with increasing mole fraction. This leveling begins when the alcohol mole fraction becomes larger than about 0.1, showing that the alcohol/water mixtures with a larger mole fraction present similar behavior in the evaporation with the pure alcohols.

Mass transfer coefficients for the evaporation of the alcohol component at a constant mole fraction and various temperatures, diffusion coefficients of the alcohol's vapor into helium at these temperatures as well as energies of activation E_a for the evaporation process, are compiled in Table 5.4. Two conclusions can be drawn from this Table. First, that the K_G values increase with increasing temperature in accord with the Arrhenius equation

Table 5.3 Mass Transfer Coefficients for the Evaporation of the Alcohol Component at Various Mole Fractions, X_A, from Alcohol/Water Mixtures, and Diffusion Coefficients of the Alcohol's Vapor into Helium at 1 atm Pressure

					10^3 D(cm^2 s^{-1})		
Alcohol	T (K)	$\dot{V}$ (cm^3 s^{-1})	10^2 X_A	10^4 K_G(cm s^{-1})	Ref. (11)	Calcd	Accuracy (%)
Methanol	323.5	0.607	2.3	134	683	688	0.7
			4.7	204	688	688	0
			10.0	239	737	688	6.7
			30.8	254	718	688	4.2
			64.0	261	708	688	2.8
			100.0	339	688	688	0
Ethanol	335.9	0.611	1.6	127	569	590	3.7
			3.3	167	605	590	2.5
			7.2	227	593	590	0.5
			24.0	266	607	590	2.8
			100.0	278	610	590	3.3
Propan-1-ol	342.4	0.681	0.49	117	470	467	0.6
			0.99	143	499	467	6.4
			2.6	174	512	467	8.8
			5.7	220	486	467	3.9
			7.4	229	480	467	2.7
			26.5	238	449	467	4.0
			48.9	264	472	467	1.1
			100.0	292	480	467	2.7

Source: Ref. (11).

Table 5.4 Mass Transfer Coefficients for the Evaporation of the Alcohol Component from Alcohol/Water Mixtures at Constant Alcohol Mole Fraction and Various Temperatures, Energies of Activation, E_a, for the Evaporation Process, and Diffusion Coefficients of the Alcohol's Vapor into Helium at 1 atm Pressure

						10^3 D(cm^2 s^{-1})		
Alcohol	10^2 X_A	$\dot{V}$ (cm^3 s^{-1})	T (K)	$10^4 \cdot K_G$ (cm s^{-1})	E_a (kJ mol^{-1})	Ref. (11)	Calcd	Accuracy (%)
Methanol	10.0	0.598	303.2	99	31.0	607	617	1.6
			312.9	134		624	654	4.8
			323.5	239		737	608	6.6
			324.2	227		674	694	3.0
			328.1	236		672	712	6.0
			333.5	288		721	729	1.1
Methanol	30.8	0.598	298.0	169	12.9	464	598	28.9
			312.6	228		537	650	21.0
			323.5	254		718	688	4.2
			329.0	279		747	713	4.6
			334.5	304		742	734	1.1
Ethanol	7.2	0.601	316.3	149	20.2	527	533	1.1
			326.0	171		546	561	2.7
			330.8	215		572	575	0.5
			335.9	227		593	590	0.5
			344.6	274		615	616	0.2
Propan-1-ol	5.7	0.632	322.7	118	35.3	400	421	5.3
			330.2	152		441	438	0.7
			336.0	246		473	452	4.4
			342.4	220		486	467	3.9
			347.2	312		556	478	14.0

Source: Ref. (11).

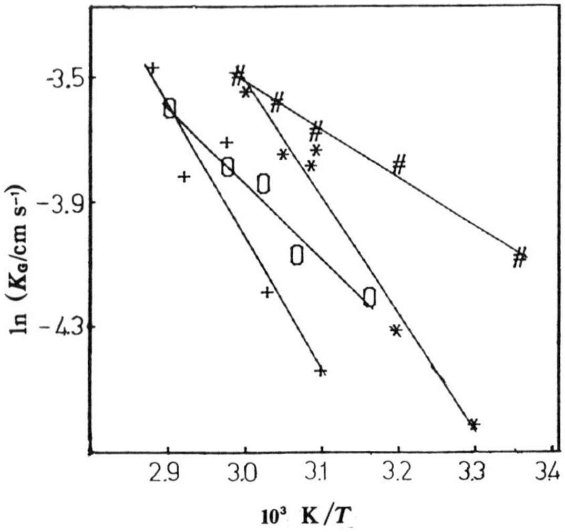

Fig. 5.10 Temperature dependence of K_G for the evaporation of the alcohol from alcohol/water mixtures: *, methanol/water (X_A = 0.100); #, methanol/water (X_A = 0.308); 0, ethanol/water (X_A = 0.072); +, propan-1-ol/water (X_A = 0.057) (11).

(cf. Fig. 5.10). This is expected, since the K_G values are rate coefficients for evaporation. Second, that the activation energy for the evaporation of the same alcohol (methanol) depends strongly on its mole fraction. All E_a values, as expected, are smaller than 40 kJ mol^{-1}, since the evaporation is a physical phenomenon.

Table 5.5 lists the values of K_G and D found at constant temperature and alcohol mole fraction, but with various carrier gas flowrates. As before, both these parameters are independent of the flowrate, since K_G is a rate coefficient and the D values of Table 5.5 are all referred to 1 atm pressure. The fact that the K_G values are independent of the carrier gas flowrate supports the boundary condition (5-33), again based on a departure from equilibrium at the liquid-gas interphase.

Table 5.5 Mass Transfer Coefficients for the Evaporation of the Alcohol Component from Alcohol/Water Mixtures at Constant Temperature and Mole Fraction, but Various Carrier Gas Flowrates. Also, Diffusion Coefficients of the Alcohol's Vapor into Helium at 1 atm Pressure.

					10^3 D (cm^2 s^{-1})		
Alcohol	T (K)	10^2 X_A	$\dot{V}$ (cm^3 s^{-1})	10^4 K_G (cm s^{-1})	Ref. (11)	Calcd	Accuracy (%)
Methanol	324.2	10.0	0.286	217	680	688	1.2
			0.408	222	698	688	1.4
			0.598	227	674	688	2.1
			0.812	225	682	688	0.9
			0.925	218	676	688	1.8
Ethanol	335.9	7.2	0.250	195	536	590	10.0
			0.398	230	572	590	3.1
			0.601	227	593	590	0.5
			0.746	227	583	590	1.2
			0.939	231	559	590	5.5
Propan-1-ol	342.4	5.7	0.278	217	483	467	3.3
			0.439	216	532	467	12.2
			0.632	220	486	467	3.9
			0.822	216	508	467	8.0
			0.957	212	464	467	0.6

Source: Ref. (11).

D. Activity Coefficients

Rearrangement of Eq. (5-42) gives the equilibrium concentration, c_0, of the solute vapor over a liquid solution, e.g., of the alcohol vapor over an alcohol-water mixture, as

$$c_0 = \frac{vh_\infty}{2}\left(\frac{L}{D} + \frac{1}{K_G}\right) \tag{5-46}$$

Since the values of D and K_G can be determined as described in detail previously, all quantities on the right-hand side of Eq. (5-46) are measurable, thus yielding the value of concentration c_0 in equilibrium with the liquid bulk phase.

If experiments are performed with liquid mixtures (in this instance alcohol + water) giving c_0, and with pure solutes (alcohols) leading to $c_0^\bullet$, both by using Eq. (5-46), then the ratio $c_0/c_0^\bullet$ is equal to $P/P^\bullet$, and this gives the activity, a_A, of the alcohol in the liquid mixture, assuming that the deviation of the solute vapor from ideal behavior is small. Then we can write

$$c_0/c_0^\bullet = a_A = \gamma_A X_A \tag{5-47}$$

and calculate the activity coefficient, γ_A, of the alcohol in the liquid mixture from the known vlaues of c_0, $c_0^\bullet$, and the mole fraction X_A.

From the activity coefficients measured, the excess chemical potential $\mu_A{}^E$ for the solute alcohol is calculated by the well-known relation

$$\mu_A{}^E = RT \ln \gamma_A \tag{5-48}$$

and from the variation of γ_A with temperature, the excess partial molar heat of mixing $H_A{}^E$ and the excess partial molar entropy of mixing $S_A{}^E$ for the solute alcohol can be found by means of the equation

$$\ln \gamma_A = \frac{H_A^E}{RT} - \frac{S_A^E}{R} \qquad (5\text{-}49)$$

which is derived from the classic relation $\mu_A^E = H_A^E - TS_A^E$ and Eq. (5-48). If a narrow temperature range is used, the above excess partial molar quantities can be assumed to remain constant, and thus plotting $\ln \gamma_A$ versus $1/T$, H_A^E is found from the slope and S_A^E from the intercept of the resulting linear plot.

The activity coefficients for the alcohol component in the binary mixtures methanol/water, ethanol/water, and propan-1-ol/water were determined from Eqs. (5-46) and (5-47), as described above (12). Table 5.6 compiles these activity coefficients for various alcohol mole fractions.

In the systems ethanol/water and propan-1-ol/water, the alcohol's activity coefficient increases with decreasing mole fraction. The positive deviation from Raoult's law in these systems has also been observed by other workers (13–15). In the case of the methanol/water system, the alcohol's activity coefficient at various X_A values shows an irregular dispersion above unity, indicating that under these experimental conditions the system approaches ideal behavior.

The activity coefficients at infinite dilution γ_A^∞ ($X_A \rightarrow 0$) can be estimated by various empirical or semiempirical equations. The most useful of these for primary alcohols/water systems is (16)

$$\log \gamma_A^\infty = a + \varepsilon N + \frac{y}{N} \qquad (5\text{-}50)$$

where N is the number of the solute's carbon atoms, and a, ε, and y are constants depending on the solute/solvent system and the temperature. Values for these constants are found in Ref. (16).

Using Eq. (5-50) one finds the activity coefficients at infinite dilution given in the last column of Table 5.6. These should

Table 5.6 Activity Coefficients for the Alcohol Component in Binary Liquid Mixtures of Alcohol and Water, at Constant Temperature and Various Alcohol Mole Fractions[a].

Alcohol	T (K)	$\dot{V}$ (cm^3 s^{-1})	10^2 X_A	γ_A	γ_A^∞
Methanol	323.5	0.607	2.3	1.33	1.814
			4.7	1.37	
			10.0	1.51	
			30.8	1.11	
			64.0	1.16	
Ethanol	335.9	0.611	1.6	6.66	4.436
			3.3	5.17	
			7.2	3.65	
			24.0	2.20	
Propan-1-ol	342.4	0.681	0.49	23.3	14.412
			0.99	19.3	
			2.6	17.1	
			5.7	13.2	
			7.4	10.1	
			26.5	3.43	
			48.9	1.88	

[a]Values at infinite dilution γ_A^∞ given in the last column have been calculated by means of Eq. (5-50).
Source: Ref. (12).

be compared with the experimental values corresponding to the smallest mole fraction of each alcohol. The differences between the two values seem rather large, but compared to the differences found in the literature between various workers and different techniques, are not at all large.

The variation of activity coefficients with temperature has been studied for all three alcohol/water systems (12), and the values found (at fairly dilute solutions) were plotted according to Eq. (5-49) as ln γ_A versus 1/T. The values of excess partial

Table 5.7 Excess Partial Molar Thermodynamic Functions of Mixing for Alcohols in Water[a]

Alcohol	X_A	T (K)	μ_A^E (kJ mol^{-1})	H_A^E (kJ mol^{-1})	S_A^E ($JK^{-1}mol^{-1}$)
Methanol	0.100	303.2	2.09		
		312.9	1.27		
		323.5	1.12	11 ± 3	30 ± 10
		324.2	1.48		
		328.1	1.30		
		333.5	0.94		
Ethanol	0.072	316.3	3.95		
		326.0	3.39		
		330.8	3.27	10 ± 3	20 ± 10
		335.9	3.51		
		344.6	3.30		
Propan-1-ol	0.057	322.7	5.02		
		330.2	5.66	-22 ± 8	-83 ± 23
		342.4	7.33		
		350.9	7.07		

[a]± Values given with H_A^E and S_A^E are standard errors, calculated by regression analysis.
Source: Ref. (12).

molar enthalpy and entropy of mixing, calculated from the slopes and intercepts of these plots, are given in Table 5.7 together with the excess chemical potential of mixing at various temperatures.

The simple relationship

$$T \ln \gamma_A = \text{constant} \tag{5-51}$$

which sometimes holds for mixtures of similar components, is not valid in this instance as expected for organic compounds/water mixtures with extensive hydrogen bonding. A simple explanation is provided by Eq. (5-49), which leads to Eq. (5-51) when $S_A{}^E$ is zero and $H_A{}^E$ remains constant with temperature. The results in Table 5.7 show that in this instance, there is a large value of excess entropy of mixing in all alcohols, and that is why Eq. (5-51) does not hold.

As a general conclusion from the results of Table 5.7, one can say that an accurate study of the variation of γ_A with temperature is very difficult, if not impossible, with classic chromatographic techniques, and hence these techniques will never probably provide accurate values of excess partial molar enthalpies of mixing as the direct calorimetric methods. One of the reasons is that careful attention must be paid to accurate measurement of retention volumes and corrections for gas-phase imperfection (17). The latter are larger when the carrier gas is helium, and this is contrary to the assumption often made that choice of helium as a carrier gas generally minimizes all the corrections. No such corrections are necessary with the present method, which is more close to a static technique than to a gas chromatographic one. No real chromatography is performed. Only gas chromatography instrumentation is used to sample the space at x = l' (cf. Fig. 5.4).

IV. MASS TRANSFER ACROSS PHASE BOUNDARIES NEAR THE JUNCTION OF THE DIFFUSION AND THE SAMPLING COLUMN

It was pointed out in the introduction of this chapter that a solid or liquid may be placed near the closed end of the diffusion column, or near the junction at x = l' of the diffusion and the sampling column (cf. Fig. 4.2), or it may fill the entire column length, L. So far, only the first of these three arrangements was considered in sections II and III. In the present section, we shall analyze briefly and generally the mass transfer phenomena with the second arrangement above, represented by Fig. 5-11. Then we shall discuss a specific case.

A. Theoretical Analysis

In the region y of the diffusion column (empty) the diffusion equation

$$\frac{\partial c_y}{\partial t_0} = D \frac{\partial^2 c_y}{\partial y^2} \tag{5-52}$$

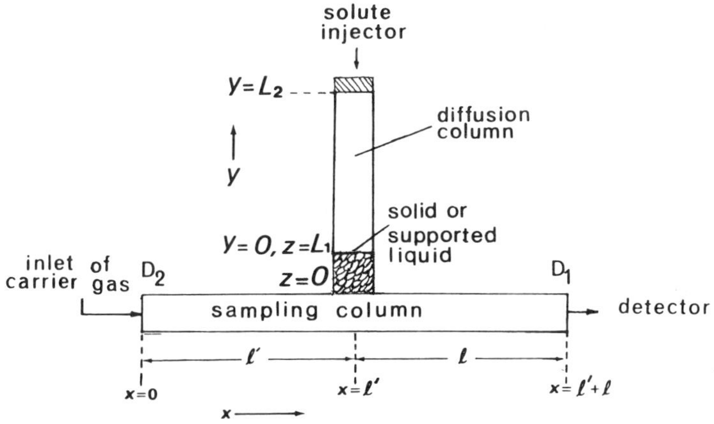

Fig. 5.11 Sampling cell for studying mass transfer phenomena across a phase boundary near the junction of the diffusion and the sampling column.

must be solved under the initial condition

$$c_y(y,0) = \frac{m}{a_G}\,\delta(y - L_2) \tag{5-53}$$

if the solute is introduced as a pulse at the injection point $y = L_2$, and the boundary condition

$$(\partial c_y/\partial y)_{y=L_2} = 0 \tag{5-54}$$

The solution at $y = 0$ is

$$C'_y(0) = \frac{m}{a_G D \cosh qL_2} - C_y(0) q \tanh qL_2 \tag{5-55}$$

where $C_y(0)$ is the t_0 Laplace transfomed function c_y at $y = 0$, $C'_y(0)$ the derivative $(dC_y/d_y)_{y=0}$, and q is given by the relation (5-16) as

$$q^2 = \frac{p_0}{D} \tag{5-56}$$

In the region z of the diffusion column (filled) the mass balance equation is

$$\frac{\partial c_z}{\partial t_0} = D\frac{\partial^2 c_z}{\partial z^2} - K_s \frac{A_s}{V'_G}(c_s^* - c_s) \tag{5-57}$$

where c_z is the gaseous concentration of the solute in the section z, V'_G the gaseous volume of the length L_1, K_s the overall mass transfer coefficient between the gas and the solid (or supported liquid phase), A_s its total surface area, and c_s^*, c_s equilibrium and nonequilibrium concentration, respectively, of solute in the solid (or supported liquid) phase.

The rate of change of c_s is given by the relation

$$\frac{\partial c_s}{\partial t_0} = K_s \frac{A_s}{V_s} (c_s^* - c_s) \tag{5-58}$$

where V_s is the total volume of the solid (or the supported liquid).

The system of partial differential equations (5-57) and (5-58) is solved by applying first Laplace transformations with respect to t_0 (parameter p_0) under the initial condition

$$c_z(z,0) = c_s(z,0) = 0 \tag{5-59}$$

and eliminating c_s between the transformed equations:

$$\frac{d^2 C_z}{dz^2} - q_z^2 C_z = 0 \tag{5-60}$$

where C_z is the t_0-transformed function $c_z(z,t_0)$ and q_z is given by

$$q_z^2 = \frac{1}{D}\left[p_0 + \frac{(K_s K A_s / V_G')p_0}{p_0 + K_s A_s / V_s}\right] \tag{5-61}$$

K being the partition coefficient in the linear isotherm

$$K = \frac{c_s^*}{c_z} \tag{5-62}$$

The solution of Eq. (5-60) using further transformation with respect to z is:

$$C_z = C_z(0) \cosh q_z z + \frac{C_z'(0)}{q_z} \sinh q_z z \tag{5-63}$$

where $C_z(0)$ and $C'_z(0)$ are the t_0-transformed functions $c_z(0)$ and $(dc_z/dz)_{z=0}$, respectively.

Eq. (5-63) is subject to two sets of boundary conditions: one at z = 0 and one at z = L_1. The first set is

$$C_z(0) = C(l',p_0)$$

$$a'_G D \left(\frac{\partial C_z}{\partial z}\right)_{z=0} = a_G v C(l',p_0) \quad (5\text{-}64)$$

a'_G and a_G being the void cross-sectional areas in the filled section L_1 and in the sampling column (or in the empty section L_2), respectively. Using this, Eq. (5-63) becomes

$$C_z = C(l',p_0)\left(\cosh q_z z + \frac{a_G v}{a'_G D q_z} \sinh q_z z\right) \quad (5\text{-}65)$$

The second set of boundary conditions at z = L_1 links together this relation of region z with Eq. (5-55) of region y. These conditions are

$$C_z(L_1) = C_y(0)$$

$$a'_G \left(\frac{\partial C_z}{\partial z}\right)_{z=L_1} = a_G \left(\frac{\partial C_y}{\partial y}\right)_{y=0} \quad (5\text{-}66)$$

and when used, they lead to the equation

$$C(l',p_0) = \frac{m}{Dq_z \cosh qL_2}\left[a'_G \sinh q_z L_1 + \frac{a_G v}{Dq_z} \cosh q_z L_1 + \frac{a_G q}{a'_G q_z} \tanh qL_2 \left(a'_G \cosh q_z L_1 + \frac{a_G v}{Dq_z} \sinh q_z L_1\right)\right]^{-1}$$

(5-67)

The usual approximation $v/Dq_z \gg 1$ permits the omission of the terms $a'_G \sinh q_zL_1$ and $a'_G \cosh q_zL_1$, thus simplifying Eq. (5-67) to

$$C(l',p_0) = \frac{m}{\dot{V}}\left(\cosh q_zL_1 \cdot \cosh qL_2 + \frac{a_Gq}{a'_Gq_z}\sinh q_zL_1 \cdot \sinh qL_2\right)^{-1} \qquad (5\text{-}68)$$

or by using the definitions $2\cosh x = \exp(x) + \exp(-x)$ and $2 \sinh x = \exp(x) - \exp(-x)$ and collecting similar terms

$$C(l',p_0) = \frac{2m}{\dot{V}}\left[\left(1 + \frac{a_Gq}{a'_Gq_z}\right)\cosh(q_zL_1 + qL_2) + \left(1 - \frac{a_Gq}{a'_Gq_z}\right)\cosh(q_zL_1 - qL_2)\right]^{-1} \qquad (5\text{-}69)$$

Expanding the cosh x in Maclaurin series and retaining the first two non-zero terms $1 + (x^2/2)$, one obtains, after rearrangement and by using the definitions (5-56) and (5-61):

$$C(l',p_0) = \frac{2mD}{\dot{V}L_1^2}\left(p_0 + \frac{K_sA_s}{V_s}\right)\left\{\left(1 + \frac{L_2^2}{L_1^2} + 2\frac{V_G}{V'_G}\right)p_0^2 + \left[\frac{2D}{L_1^2} + \frac{K_sKA_s}{V'_G} + \frac{K_sA_s}{V_s}\left(1 + \frac{L_2^2}{L_1^2} + 2\frac{V_G}{V'_G}\right)\right]p_0 + \frac{2D}{L_1^2}\cdot\frac{K_sA_s}{V_s}\right\}^{-1} \qquad (5\text{-}70)$$

where V_G and V'_G are the gaseous volumes in the empty (L_2) and the filled (L_1) section of the column, respectively.

Inversing (5-70), we find the same form of Eq. (5-28):

$$c(l',t_0) = \frac{N_5}{2}\left[\left(1 + \frac{Z_2}{Y_2}\right)\exp\left(-\frac{X_2 + Y_2}{2}t_0\right) + \left(1 - \frac{Z_2}{Y_2}\right)\exp\left(-\frac{X_2 - Y_2}{2}t_0\right)\right] \tag{5-71}$$

where

$$X_2 = \frac{2D/L_1^2 + K_sKA_s/V'_G}{1 + L_2^2/L_1^2 + 2V_G/V'_G} + \frac{K_sA_s}{V_s} \tag{5-72}$$

$$Y_2 = \left[X_2^2 - \frac{8DK_sA_s/L_1^2V_s}{1 + L_2^2/L_1^2 + 2V_G/V'_G}\right]^{1/2} \tag{5-73}$$

$$Z_2 = \frac{2D/L_1^2 + K_sKA_s/V'_G}{1 + L_2^2/L_1^2 + 2V_G/V'_G} - \frac{K_sA_s}{V_s} \tag{5-74}$$

and

$$N_5 = \frac{2mD}{\dot{V}L_1^2(1 + L_2^2/L_1^2 + 2V_G/V'_G)} \tag{5-75}$$

Eq. (5-71) can be used to analyze experimental data in the same way described for Eq. (5-28).

If the mass transfer rate in the solid is high, p_0 can be omitted in the denominator of Eq. (5-61), which becomes

$$q_z^2 = \frac{p_0}{D}\left(1 + K\frac{V_s}{V'_G}\right) = \frac{p_0}{D}(1 + k) \tag{5-76}$$

where k is the partition ratio as given by Eq. (1-5). Then, instead of Eq. (5-70), the following expression is obtained:

$$C(l',p_0) = \frac{2mD}{\dot{V}L_1^2}\left[\left(1 + k + \frac{L_2^2}{L_1^2} + 2\frac{V_G}{V'_G}\right)p_0 + \frac{2D}{L_1^2}\right]^{-1} \tag{5-77}$$

which on inverting gives

$$c(l',t_0) = N_6 \exp\left[-\frac{2D/L_1^2}{1 + k + L_2^2/L_1^2 + 2V_G/V'_G}\,t_0\right] \tag{5-78}$$

where N_6 is a constant similar to N_5.

Eq. (5-78) is a simple exponential function which can be used to calculate k from the slope of ln h vs. t_0 plot of experimental data. Another way to find k is to apply an approximation used before for small times (18): Replace both sinh x and cosh x in Eq. (5-68) by exp (x)/2. We then obtain

$$c(l',t_0) = \frac{N_7}{t_0^{3/2}} \exp\left(-\frac{L_{eff}^2}{4Dt_0}\right) \tag{5-79}$$

where N_7 is a constant and L_{eff} an effective length of the column given by

$$L_{eff} = L_2 + L_1(1 + k)^{1/2} \tag{5-80}$$

Eq. (5-79) coincides with Eq. (45) of Ref. 19, derived in a somewhat different way (18), assuming reversible adsorption of the solute on the solid and for different temperatures of the two sections, L_1 and L_2. It resembles Eq. (4-19) for pure diffusion and can be applied in a similar way as that equation, i.e., by plotting ln $ht_0^{3/2}$ vs. $1/t_0$. From the slope and the known lengths, L_1 and L_2, one can determine k.

Finally, from the value of k the adsorption equilibrium constant, K, can be found using the relation $K = k\varepsilon/(1 - \varepsilon)$, ε being the void fraction in the filled column section L_1. This is determined by a standard method (20).

B. Some Representative Results

Partition ratios and adsorption equilibrium constants at various temperatures, determined by the method outlined above, have been reported elsewhere (18). They refer to the adsorption of methane, ethane, n-butane, ethene, propene, and but-1-ene on aluminum oxide from helium.

The k values for methane are worth mentioning because they are small, but not zero as usually assumed by chromatographers who use this gas to determine gas hold-up time in gas-solid chromatography (GSC) with a flame ionization detector. At 323°K, for example, the k value of CH_4 is 0.792, which means that a gas hold-up time 79% greater than the actual value would be determined.

From the variation of the adsorption equilibrium constants with temperature, the differential enthalpies (heats) and entropies of adsorption can be determined using the relation (18):

$$\ln \frac{K_a}{T} = \ln R + \frac{\Delta' S^\ominus}{R} - \frac{\Delta H^\ominus}{R} \cdot \frac{1}{T} \tag{5-81}$$

A plot of ln (K/T) versus 1/T will give $\Delta H^\ominus$ from the slope and $\Delta S^\ominus$ from the intercept, it being understood that the range of T is narrow enough for $\Delta H^\ominus$ and $\Delta S^\ominus$ to be regarded as independent of temperature. Some results are presented in Table 5.8. The values of $\Delta S^\ominus$ given should be regarded as relative. They are referred to an hypothetical adsorbed standard state of 1 mol/cm^3 of solid, and a partial pressure of 1 atm as the gas standard state.

Table 5.8 Enthalpies and Entropies of Adsorption of Various Hydrocarbons on Al_2O_3 from Helium

Substance	$-\Delta H^{\ominus}$ (kJ mol^{-1})	$-\Delta S^{\ominus}$ (J K^{-1} mol^{-1})
Methane	13 ± 2	143 ± 6
Ethane	14.1 ± 0.4	129 ± 1
n-Butane	24 ± 3	144 ± 7
Ethene	23 ± 1	149 ± 3
Propene	38 ± 4	189 ± 12
But-1-ene	43 ± 2	190 ± 6

Source: Ref. (18).

The merits of the above method for determining adsorption equilibrium constants are reflected in the differential heats of adsorption found, which corresponds to isosteric of zero coverage more closely than values determined from conventional gas chromatographic data by plotting ln V_N or ln (V_N/T) versus $1/T$. This is because the total amount of solute injected into the system is of the order of 10^{-5} mol and, owing to its diffusional distribution along the column $L_1 + L_2$, only a small fraction of this is located over the adsorbent bed, and this is almost uniformly distributed all the time. By contrast, in usual chromatographic elution the whole amount of solute moves along the column as a relatively narrow band, with much higher local concentrations at a certain region of the adsorbent and zero concentration at others.

Another aspect of the data in Table 5.8 is the absence of a compensation effect; i.e., a linear dependence of $\Delta S^{\ominus}$ on $\Delta H^{\ominus}$, which is almost always observed with ΔS and ΔH values determined chromatographically. An explanation of the compensation effect (21) is that it is due to the chromatographic process itself combined with the heterogeneity of the surface. The absence of a compensation effect in this instance can thus be due to the fact that there is no chromatography inside the diffusion column. The enthalpies and entropies of adsorption in Table 5.8 have been determined by gas chromatography instrumentation without "chromatography."

V. MASS TRANSFER ACROSS PHASE BOUNDARIES EXTENDED TO ALL THE DIFFUSION COLUMN

As a final case, we examine the arrangement of having the entire length of the diffusion column filled with a solid (or supported liquid) phase.

A. Theoretical Analysis

The necessary length coordinates are those of Fig. 4.2, and the mass balance equation is given by (5-57), with all symbols having the same meaning as before.

The rate of change of c_s under nonsteady-state conditions is again given by Eq. (5-58), but to make the discussion a little more general, let us assume that the transfer of solute on the solid surface is followed by a first-order chemical reaction with rate constant k_2:

$$\frac{\partial c_s}{\partial t_0} = \frac{K_s A_s}{V_s}(c_s^* - c_s) - k_2 c_s \tag{5-82}$$

The initial conditions here are chosen as

$$c_z(z,0) = c_0 \qquad c_s(z,0) = 0 \tag{5-83}$$

i.e., the entire gaseous column length is filled with a constant concentration, c_0, initially, whereas no solute has been transferred as yet to the solid.

The boundary conditions are

$$\left(\frac{\partial c_z}{\partial z}\right)_{z=L} = 0 \tag{5-84}$$

at z = L, since there is no flux across this boundary, and

$$c_z(0,t_0) = c(l',t_0)$$

$$a'_G D\left(\frac{\partial c_z}{\partial z}\right)_{z=0} = a_G v c(l',t_0) \qquad (5\text{-}85)$$

at the boundary z = 0, a'_G and a_G being the corss-sectinal areas of column L and l' + l, respectively.

Following the same procedure as in the previous section of this chapter, we find the solution, with respect to the independent variable z, of the system of Eqs. (5-57) and (5-82), under the initial condition (5-83) and subject to the boundary conditions (5-84) and (5-85), as

$$C(l',p_0 0 = \frac{c_0}{Dq_1^2} \cdot \frac{\sinh q_1 L}{\sinh q_1 L + (a_G v/a'_G Dq_1) \cosh q_1 L} \qquad (5\text{-}86)$$

where $C(l',p_0)$ is the t_0 Laplace-transformed fucntion $c(l',t_0)$, and q_1 is given by the relation

$$q_1{}^2 = \frac{1}{D}\left[p_0 + \frac{k_1\ (p_0 + k_2)}{p_0 + k_{-1} + k_2}\right] \qquad (5\text{-}87)$$

with k_1 and k_{-1} denoting the groups of constants

$$k_1 = \frac{K_s K A_s}{V'_G} \qquad (5\text{-}88)$$

$$k_{-1} = \frac{K_s A_s}{V_s} \qquad (5\text{-}89)$$

Omitting sinh $q_1 L$ from the denominator of Eq. (5-86) as usual, we obtain

$$C(l',p_0) = \frac{c_0 a'_G D}{\dot{V}} \cdot \frac{\tanh q_1 L}{Dq_1} \tag{5-90}$$

Eq. (5-25) shows that $(\tanh qL)/Dq$ is approximated by $2/L(p_0 + \pi^2 D/4L^2)$ when $q^2 = p_0/D$ according to Eq. (5-16). Now that q_1^2 is given by Eq. (5-87), we consider $k_1(p_0 + k_2)/(p_0 + k_{-1} + k_2)$ as a perturbation added to p_0. We therefore replace p_0 in the denominator of Eq. (5-25) by the expression in brackets [ ] of Eq. (5-87):

$$\frac{\tanh q_1 L}{Dq_1} \simeq \frac{2}{L}\left[p_0 + \frac{k_1(p_0 + k_2)}{p_0 + k_{-1} + k_2} + \frac{\pi^2 D}{4L^2}\right]^{-1} \tag{5-91}$$

Finding now the p_0 inverse transform of Eq. (5-90) with the help of the approximation (5-91), we have, after rearrangement

$$c(l',t_0) = \frac{N_8}{2}\left[\left(1 + \frac{Z_3}{Y_3}\right)\exp\left(-\frac{X_3 + Y_3}{2}t_0\right) + \left(1 - \frac{Z_3}{Y_3}\right)\exp\left(-\frac{X_3 - Y_3}{2}t_0\right)\right] \tag{5-92}$$

where

$$X_3 = \frac{\pi^2 D}{4L^2} + k_1 + k_{-1} + k_2 \tag{5-93}$$

$$Y_3 = \left[X_3^2 - \frac{\pi^2 D}{L^2}(k_{-1} + k_2) - 4k_1 k_2\right]^{1/2} \tag{5-94}$$

$$Z_3 = \frac{\pi^2 D}{4L^2} + k_1 - k_{-1} - k_2 \tag{5-95}$$

$$N_8 = \frac{2c_0 a'_G D}{\dot{V}L} \tag{5-96}$$

Two limiting cases of Eq. (5-92) are interesting. They arise from the relative magnitudes of the terms of Eq. (5-93). First, when k_1 is small compared to the diffusion or the other terms, i.e., when the mass transfer from the gas phase to the solid is slow, Eq. (5-92) reduces to the relation

$$c(l',t_0) = N_8 \exp\left(\frac{-\pi^2 D t_0}{4L^2}\right) \tag{5-97}$$

applicable to simple diffusion, as expected. Second, when k_{-1} is small compared with the other terms, meaning that mass transfer from the surface to the bulk of the solid phase (when $c_s^* > c_s$) or vice versa (when $c_s^* < c_s$) is slow, Eq. (5-92) becomes

$$c(l',t_0) = N_8 \exp\left[-\left(\frac{\pi^2 D}{4L^2} + k_1\right) t_0\right] \tag{5-98}$$

i.e., it reduces to a simple exponential function of the same form as Eq. (5-97), but with a different exponent. These two limiting cases work irrespective of the rate of the chemical reaction in the solid, as expressed by k_2.

In order to see whether experimental data conform to either Eq. (5-97) or Eq. (5-98), it is only necessary to use Eq. (5-1) with $c(l',t_0)$ substituted by the right-hand side of these equations, and taking logarithms. In the case of Eq. (5-97), this leads to Eq. (4-39) of simple diffusion, whereas from Eq. (5-98) the following expression is obtained:

$$\ln h = \ln (2N_8) - \left(\frac{\pi^2 D}{4L^2} + k_1\right) t_0 \tag{5-99}$$

The common criterion is that after the first few points, owing to the second and the higher terms in the approximation (5-25), a straight line is obtained when plotting ln h vs. t_0. The absolute value of the slope of this line is either $\pi^2D/4L^2$, corresponding to the limiting case $k_1 \simeq 0$; or $(\pi^2D/4L^2 + k_1)$; i.e., greater than before, corresponding to the second limiting case $k_{-1} \simeq 0$. From the latter slope k_1 is determined.

If the plot of ln h vs. t_0 is not linear, it corresponds to the general form of Eq. (5-92), which predicts a curve described by the sum of two exponential functions. These functions have different exponential coefficients, $(X_3 + Y_3)/2$ and $(X_3 - Y_3)/2$, and if the difference of them is big enough, X_3 and Y_3 can be computed from the slopes of the plot. Then, using also the known value of $\pi^2D/4L^2$ in Eqs. (5-93), (5-94), and (5-95) combinations of the constants k_1, k_{-1} and k_2 can be found.

B. The Kinetics of Action of Sulfur Dioxide on Marble

By using a gas chromatograph equipped with a flame photometric detector (FPD), the kinetics of sulfur dioxide action on marble was studied (22) by means of a 32.5 cm × 4 mm i.d. diffusion column filled with marble pieces. The carrier gas was dry or wet air.

First, Eq. (5-97) substituted in the relation

$$h = [2c(l',t_0)]^m \tag{5-100}$$

where m is the response factor of FPD, different from unity, gives

$$\ln h = m \ln (2N_8) - \frac{m\pi^2Dt_0}{4L^2} \tag{5-101}$$

and this provides a means to find m by injecting gaseous SO_2 into the empty diffusion column and plotting ln h vs. t_0 to find the value of $m\pi^2D/4L^2$. Then, using the theoretically calculated (6) value of D, we can find m, since L is accurately known.

Second, Eq. (5-98) must be substituted in (5-100), rather than in Eq. (5-1) for a filled diffusion column, to obtain a modified form of Eq. (5-99):

$$\ln h = m \ln (2N_8) - m\left(\frac{\pi^2 D}{4L^2} + k_1\right) t_0 \qquad (5\text{-}102)$$

An example of plotting the experimental data as ln h vs. t_0 is given in Fig. 5-12. It is seen that both kinds of plot with empty and filled diffusion columns are linear to a good approxi-

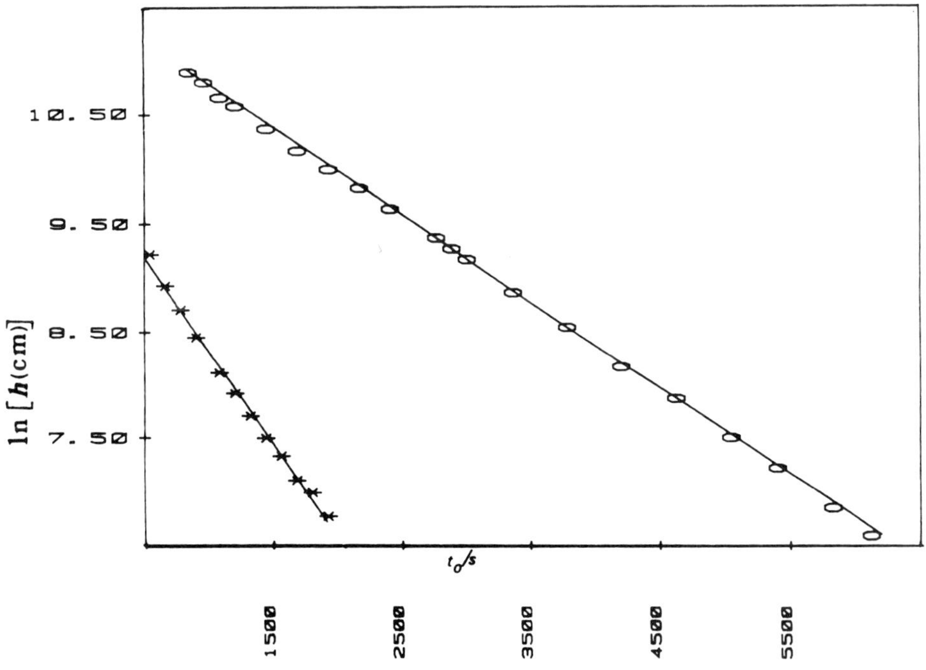

Fig. 5.12 The height of the sample peaks vs. t_0 on a semilogarithmic scale, obtained with an empty (o) and a filled (*) with marble pieces diffusion column (22).

mation, differing only in their slopes, in accord with Eq. (5-101) and (5-102). This has been observed with two marble samples of different particle size at all temperatures, and with either dry or wet air as carrier gas.

The response factor, m, determined at various temperatures by means of Eq. (5-101) was found to be 1.92 (mean value).

From the results obtained with the column filled with $CaCO_3$, it seems that the slopes of the plots of ln h vs. t_0 decrease initially as the number of SO_2 injections increases. This is reasonable since as the total quantity of SO_2 which passes over $CaCO_3$ increases, various active centers of the surface are saturated and the rate of the mass transfer to the solid decreases. After a certain number of SO_2 injections, the interaction with $CaCO_3$ is stabilized and the slope of the plots remains unchanged. Dividing each slope with the m value, one finds the value of $\pi^2D/4L^2 + k_1$ (cf. Eq. (5-102)). Tables 5.9 and 5.10 collect such values of the two marble species, obtained with dry and wet air as carrier gas. Each of these values represents the mean of at least three kinetic runs conducted under identical experimental

Table 5.9 Values of the Kinetic Parameter $\pi^2D/4L^2 + k_1$ of Eq. (5-98), and the Rate Constant k_1 of Eq. (5-88) for Marble Particles 22-44 Mesh and Surface Area 0.68 m^2g^{-1}

	$10^4(\pi^2D/4L^2 + k_1)/s^{-1}$		$10^4\ k_1/s^{-1}$
T (K)	Dry air	Wet air	Wet air
322.2	3.44	12.50	9.19
353.2	3.80	7.29	3.43
373.2	3.26	4.18	—
393.2	3.18	4.23	—
423.2	4.32	7.00	1.74

Source: Ref. (22).

Table 5.10 Values of the Kinetic Parameter $\pi^2D/4L^2 + k_1$ of Eq. (5-98), and the Rate Constant k_1 of Eq. (5-88) for Marble Particles 120-150 Mesh and Surface Area 1.15 m^2 g^{-1}

	$10^4(\pi^2D/4L^2 + k_1)/s^{-1}$		$10^4\ k_1/s^{-1}$	
T(K)	Dry air	Wet air	Dry air	Wet air
322.2	7.61	13.78	4.30	10.47
353.2	8.02	10.23	4.16	6.37
373.2	9.96	15.77	5.70	11.51
393.2	3.13	4.53	—	—
423.2	5.59	8.85	—	3.59

Source: Ref. (22).

conditions. The k_1 values of Eq. (5-88) included in Tables 5.9 and 5.10 have been found by using the calculated values of $\pi^2D/4L^2$ listed in Table 5.11.

From the experimental finding that all plots of ln h vs. t_0 are linear, we can conclude that the rate of the action of SO_2 on $CaCO_3$, under the conditions used, conforms to the limiting cases described by Eqs. (5-97) and (5-98). Comparing the values of the diffusion parameter $\pi^2D/4L^2$ of Table 5.11 with

Table 5.11 Values of the Diffusion Parameter $\pi^2D/4L^2$ of Eq. (5-98)

T (K)	D_{calcd} (cm^2s^{-1})	p (atm)	$10^4\pi^2D/4L^2/s^{-1}$
322.2	0.147	1.038	3.31
353.2	0.172	1.040	3.86
373.2	0.190	1.043	4.26
393.2	0.209	1.046	4.67
423.2	0.236	1.049	5.26

Source: Ref. (22).

the values of the kinetic parameter $\pi^2D/4L^2 + k_1$ in Table 5.9, we see that there is not a significant difference between the two parameters at the same temperature when dry air is used as carrier gas. This means that the rate constant k_1 has a very small value owing to small values of one or more of the factors on the right-hand side of Eq. (5-88). When the carrier gas (air) is saturated with water vapor, however, there is a significant difference between the kinetic and the diffusion parameter, but only at temperatures below 100°C. At these temperatures there is a non-zero rate of action of SO_2 on $CaCO_3$ and Eq. (5-98) is obeyed, meaning that mass transfer from the surface to the bulk of the solid phase or vice versa is very slow ($k_{-1} \simeq 0$). Obviously, the difference observed with wet air is due to water condensing or adsorbed on the solid surface. At temperatures near 100°C most of this water is removed from the surface of the solid and the effect of its presence diminishes. Thus an increase in temperature cannot bring about an increase in the transfer rate of SO_2 onto the solid surface.

Having Eq. (5-88) as a guide, one can increase the k_1 value at the same temperature by increasing the surface area, A_s, using a smaller particle size, thus increasing the value of k_1. The results of Table 5.10 confirm this prediction, at least qualitatively. At the higher temperatures there is another factor diminishing the rate of SO_2 action, and that is the partition coefficient, K, which is expected to decrease with increasing temperature, thus causing a decrease in k_1.

In conclusion, at relatively high temperatures and dry air the rate of action of SO_2 on $CaCO_3$ having a low surface area is negligible. At lower temperatures, increasing the water content of the air and the surface area of the solid increases the rate of action, its rate-determining step being the mass transfer from the gas phase to the surface of the solid. The rate of the mass transfer from the surface to the bulk of the solid phase (or vice versa), and the rate of the chemical reaction between SO_2 and $CaCO_3$ are very slow.

LIST OF SYMBOLS

a_A	Activity in a liquid
a_G, a'_G	Volume of gas phase per unit length of column, or cross-sectional area of void space
a_L	Free surface area of the liquid
A_s	Total surface area
c	Concentration of a solute vapor in the sampling column
c_0	Equilibrium vapor concentration, initial concentration
c_L	Concentration of the absorbed solute in the bulk liquid phase
c_L^*	Equilibrium solute concentration in the liquid phase
c_y, c_z	Concentration of a solute vapor in the diffusion column
c_s^*, c_s	Equilibrium and nonequilibrium concentration, respectively
C, C_L, C_L^*, C_z	Laplace transforms of c, c_L, c_L^*, and c_z with respect to t_0
$\bar{C}_z$	Double Laplace transform of c_z with respect to t_0 and z
D	Mutual diffusion coefficient of two gases
h	Height of a sample peak measured from the ending baseline
h_∞	Infinity peak height defined by Eq. (5-42)
$H^\ddagger$	Henry's law constant
$H_A{}^E$	Excess partial molar heat mixing
k, k'	Rate constants defined by Eqs. (5-20) and (5-21), respectively
k_1, k_{-1}	Rate constants defined by Eqs. (5-88) and (5-89), respectively
k_2	Rate constant of chemical reaction

k_G, k_L	Gas and liquid film transfer coefficients, respectively
K	Partition coefficient
K_G	Overall mass transfer coefficient in the gas phase
K_s	Overall mass transfer coefficient between gas and solid
K_L	Overall mass transfer coefficient in the liquid phase
L_{eff}	Effective diffusion length defined by Eq. (5-80)
l, l'	Lengths of two sections of the sampling column
L, L_1, L_2	Length of the diffusion column
m	Amount of solute injected, response factor of FPD
N_2, N_5, N_8	Constants defined by Eqs. (4-25), (5-75), and (5-96), respectively
p_0	Transform parameter with respect to t_0
q, q_1, q_z	Parameters defined by Eqs. (5-16), (5-87), and (5-61), respectively
s	Transform parameter with respect to z
$S_A{}^E$	Excess partial molar entropy of mixing
t_0	Time from the beginning to the last backward reversal of gas flow
v	Linear velocity of carrier gas in interparticle space of the sampling column
$\dot{V}$	Volume flowrate of carrier gas
V_G, V'_G	Gaseous volumes in the diffusion column
V_L	Volume of the liquid
V_s	Total volume of solid
x, z	Distance coordinates in column $l' + l$ or L, respectively
X_1, Y_1, Z_1	Parameters defined by Eqs. (5-29), (5-30), and (5-31), respectively
X_2, Y_2, Z_2	Parameters defined by Eqs. (5-72), (5-73), and (5-74), respectively
α	Diffusion parameter defined by Eq. (5-26)

γ_A	Activity coefficient in a liquid
γ_A^{∞}	Activity coefficient at infinite dilution
μ_A^E	Excess chemical potential

REFERENCES

1. D. A Blackadder and R. M. Nedderman, A Handbook of Unit Operations, Academic Press, New York, 1971, p. 117
2. W. G. Whitman, Chem. and Met. Eng., 29:147 (1923).
3. N. A. Katsanos and E. Dalas, J. Phys. Chem., 91:3103 (1987).
4. G. Karaiskakis and N. A. Katsanos, J. Phys. Chem., 88: 3674 (1984).
5. M. L. Boas, Mathematical Methods in the Physical Sciences, Wiley, New York, 1966, p. 407.
6. R. B. Bird, W. E. Stewart, and E. N. Lightfoot, Transport Phenomena, Wiley, New York, 1960, p. 511.
7. V. R. Maynard and E. Grushka, Adv. Chromatogr., 12:99 (1975).
8. E. N. Fuller, P. D. Schettler, and J. C. Giddings, Ind. Eng. Chem., 58:19 (1966).
9. J. Crank, The Mathematics of Diffusion, Oxford University Press, Oxford, England, 1956, p. 34.
10. Reference 6, p. 522.
11. G. Karaiskakis, P. Agathonos, A. Niotis, and N. A. Katsanos, J. Chromatogr., 364:79 (1986).
12. N. A. Katsanos, G. Karaiskakis, and P. Agathonos, J. Chromatogr., 349:369 (1986).
13. R. H. Rerry and C. H. Chilton, Chemical Engineer's Handbook, McGraw-Hill, New York, 5th ed., 1973, pp. 13-9 and 13-10.
14. R. S. Hansen and F. A. Miller, J. Phys. Chem., 58:193 (1954).
15. D. A. Shaw and T. F. Anderson, Ind. Eng. Chem. Fundam., 22:79 (1983).
16. R. C. Reid, J. M. Prausnitz, and T. K. Sherwood, The Properties of Gases and Liquids, McGraw-Hill, New York, 3rd ed., 1977, pp. 336 and 337.

17. J. R. Conder and C. L. Young, Physicochemical Measurements by Gas Chromatography, Wiley, New York, 1979, p. 15.

18. G. Karaiskakis, N. A. Katsanos, and A. Niotis, J. Chromatogr., 245:21, (1982).

19. N. A. Katsanos and G. Karaiskakis, Adv. Chromatogr., 24:125 (1984).

20. A. S. Foust, L. A. Wenzel, C. W. Clump, L. Maus, and L. B. Anderson, Principles of Unit Operations, Wiley, New York, 1960, p.474.

21. N. A. Katsanos, A. Lycourghiotis, and A. Tsiatsios, J. Chem. Soc., Faraday Trans. I, 74:575 (1978).

22. N. A. Katsanos and G. Karaiskakis, J. Chromatogr., 395: 423 (1987).

6

Reversed-Flow with a Filled Sampling Column

I. Introduction

II. Kinetics of Surface-Catalyzed Reactions

- A. General Methodology
- B. Experimental Setup

III. Reaction Kinetics with the Pulse Technique

- A. Theory
- B. Dehydration of Alcohols
- C. Deamination of Primary Amines
- D. Catalytic Cracking of Cumene

IV. Reaction Kinetics with the Diffusion Technique

- A. Theory
- B. Oxidation of Carbon Monoxide
- C. Hydrogenation of Propene
- D. Dehydrosulfurization of Thiophene
- E. Methanation of Carbon Monoxide

V. Mass Transfer Coefficients of Vapors on Solids

List of Symbols

References

I. INTRODUCTION

The applications of the reversed-flow gas chromatographic technique (RFGC) described in Chapters 4 and 5 were all with an empty sampling column. The latter was used for the mere creation of sample peaks by placing the carrier gas stream perpendicular to the diffusion stream in the column L (cf. Fig. 3.1), and reversing the direction of the carrier gas flow from time-to-time. In only one case (see Chapter 4, section II), the sampling column was filled with silica gel to act as a separation material for the components of a gas mixture (cf. Fig. 4.9). In the present chapter, applications with a sampling column containing a solid catalyst (1-6), or a chromatographic material (7-9), or both (9,10) will be examined. The diffusion column will be either absent, or empty of any solid, or partly filled with a catalyst near the junction with the sampling column.

II. KINETICS OF SURFACE-CATALYZED REACTIONS

A. General Methodology

When the chromatographic sampling equation was derived in Chapter 3, section IV, we assumed that a rate process was taking place in the region $x = l'$ of the sampling column (cf. Fig, 3.4), and produced an increase or decrease with time in the concentration $c(l',t_0)$ of solute A at this region. If this rate process is not physical in nature, as those described in the previous two chapters, but a chemical reaction occurring on the surface of a solid catalyst contained in the sampling column, the concentration(s) $c(l',t_0)$ as a function of time will generally be determined by: 1. the method used to feed the reactant(s) to the catalyst bed; 2. the rate of the chemical reaction on the catalyst, $r(t_0)$; 3. the catalyst's weight, W; and 4. the volume flow rate, $\dot{V}$, of the carrier gas in the sampling column.

There are two main methods for feeding the reactant(s) onto the catalytic bed. The first, termed pulse technique, is used when the reactant is strongly adsorbed on the active surface, and the catalyst can also function as chromatographic material to separate the products from one another and from the reactant. In this case, only the sampling column $l' + l$ is employed, filled with the catalyst, and no diffusion column is used. The reactant is injected as a pulse at a middle position of the column, as shown in Fig. 6.1. Simple flow reversals then create sample peaks like those shown in Fig. 6.2.

The previous technique cannot be used when the reactant is not retained on the catalytic bed for a sufficiently long time, but is eluted together with the product(s). It cannot also be applied with two gaseous reactants unless one of them is used in a great excess of the other as a carrier gas. In these situations, a continuous feed of reactant(s) into the catalytic bed can be arranged by using a diffusion column L (diffusion technique) and introducing the reactant or a mixture of reactants at the top of the column L (Fig. 6.3). These are then allowed to diffuse slowly onto the catalyst retained within a short section near $x = l'$. A catalytically inert chromatographic material is used for separation, either filling with it column $l + l'$, or loading with it an additional separation column connected, as shown in Fig. 6.3. This can be heated to a different temperature than the catalyst. The flow reversal at a certain time now samples all substances present at $x = l'$, and the chromatographic material separates these substances, revealing their relative concentrations. An example is given in Fig. 6.4.

B. Experimental Setup

1. Pulse Technique

The experimental setup in the pulse technique is depicted in Fig. 6.5.

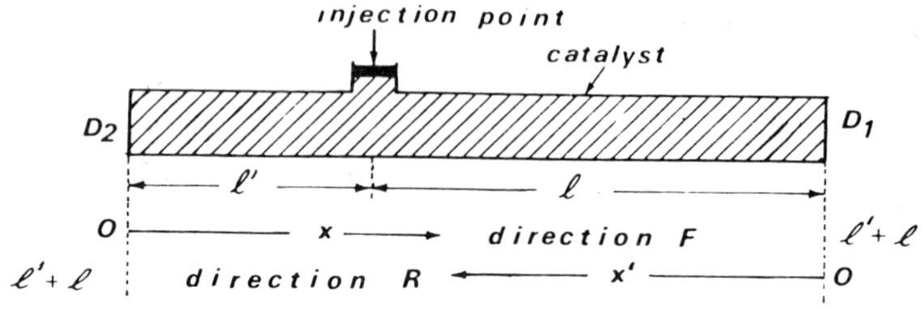

Fig. 6.1 Representation of a catalytic sampling column l' + l to study heterogeneous catalysis of a reactant strongly adsorbed on the active surface (2).

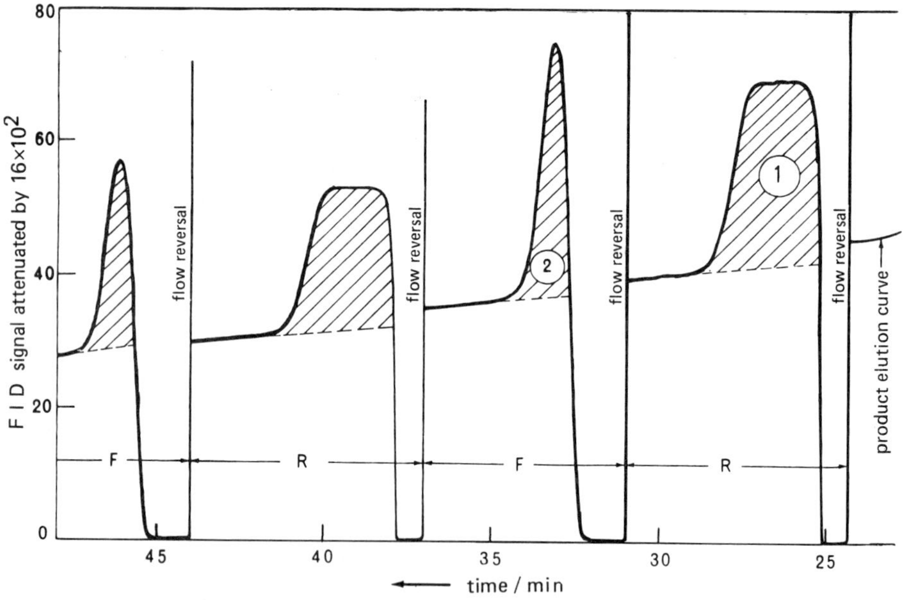

Fig. 6.2 Sample peaks obtained with a 3.8-mm i.d. glass column of lengths l' = 7 and l = 100 cm, filled with 80-100 mesh 13X molecular sieve, activated at 665°K for 21 h. The carrier gas was nitrogen ($\dot{V}$ = 0.446 $cm^3\ s^{-1}$), the column temperature 477.3°K, and the injected reactant 0.5 mm^3 propan-2-ol giving di-isopropyl ether as the main product (2).

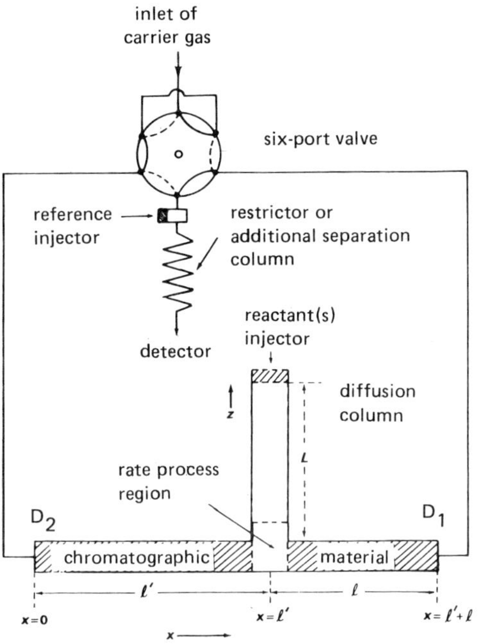

Fig. 6.3 Schematic representation of the columns and gas connections when using the diffusion technique (11).

After conditioning the catalyst by heating it in situ at an appropriate temperature, under carrier gas flowing in either direction, some preliminary injections of the reactant in both directions are made to establish constant catalytic activity. Kinetic experiments are conducted after the last reactant introduced has been exhausted to a negligible amount. A few cubic millimeters (0.5–10.0) of liquid reactant using a microsyringe, or a few cubic centimeters of gas reactant at atmospheric pressure using a gas-tight syringe, are introduced onto the column length, l, with the carrier gas flowing in direction F. The product(s) are recorded by the detector as an asymmetric elution curve, and the chemical reaction can be sampled in the usual way by reversing the direction of the carrier gas flow. Either case analyzed in Chapter 3,

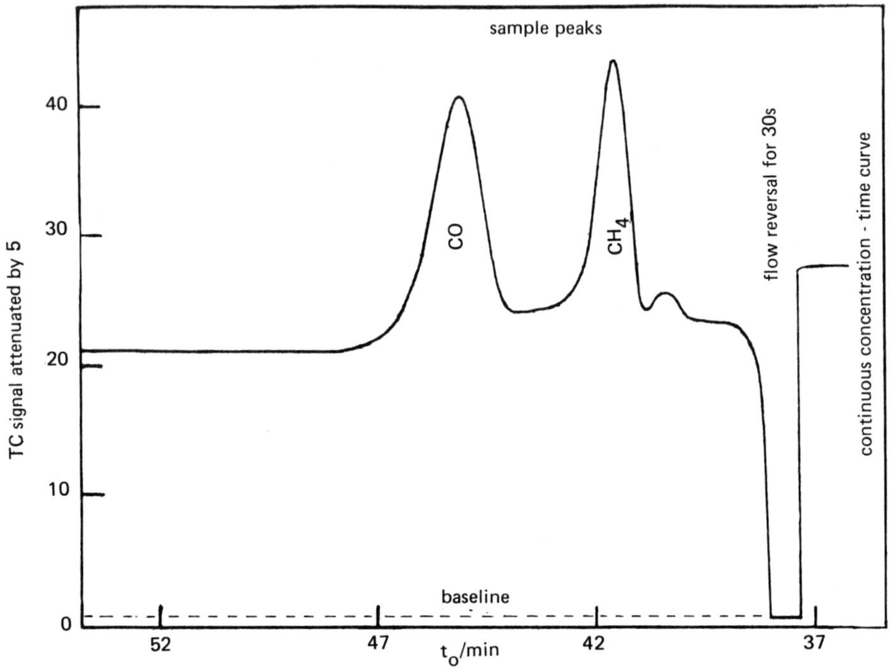

Fig. 6.4 Sample peaks of CO (reactant) and CH_4 (product) created by a 30s flow reversal of carrier gas $H_2(\dot{V}=0.75\ cm^3\ s^{-1})$, through a sampling cell (115 + 115 cm × 4 mm i.d.) filled with molecular sieve 5A and containing 53 mg of 5% Ni/Al_2O_3 catalyst placed near the junction at x = l'. The CO was introduced 37 min earlier through the solute injector of a 100-cm long (4 mm i.d.) diffusion column, and the temperature was 489°K (10).

section IV.C, i.e., $t' > t_R + t'_R$ or $t' < t_R + t'_R$, can be used, giving one sample peak after each flow reversal in the first case, or a higher (almost double) sample peak after two successive reversals in the second case.

2. Diffusion Technique

In this case, the experimental arrangement is a combination of that outlined in Figs. 6.1 and 6.5 with that used to study diffusion coefficients, as depicted in Figs. 4.1 and 4.2. Two

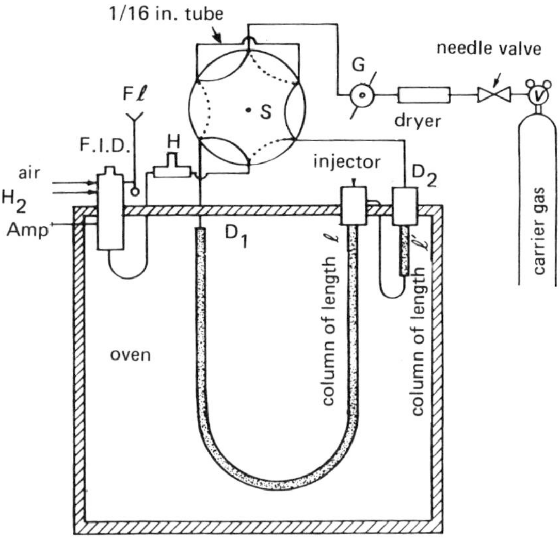

Fig. 6.5 Schematic arrangement of the experimental setup for studying the kinetics of a heterogeneously catalyzed reaction of a substance strongly adsorbed on the catalytic surface. V = two stage reducing valve and pressure regulator; G = gas flow controller for minimizing variations in the gas flow-rate; S = six-port gas sampling valve; H = restrictor; Fl = bubble flowmeter; Amp = signal to amplifier. (2).

examples are shown in Figs. 6.6 and 6.7, both of which conform to Fig. 6.3. The reactant(s) are introduced at the top of the diffusion column L, either by injecting a small volume (0.5 to 1.0 cm^3) of gas as a pulse, or a bigger volume slowly to fill column L (e.g., 10 cm^3 at atmospheric pressure), or by connecting at the top of column L a U-shaped tube containing about 0.5 cm^3 of liquid volatile reactant, as shown in Fig. 6.7.

In most cases, a conventional gas chromatograph is used, modified so that it accommodates inside its oven the reversed-flow cell, connected to the carrier gas inlet and the chromatographic detector through a four- or six-port valve. All three branches of the cell, L, l' and l, are of glass or stainless-steel chromatographic tube with i.d. 2-4 mm, and lengths 50-120 cm.

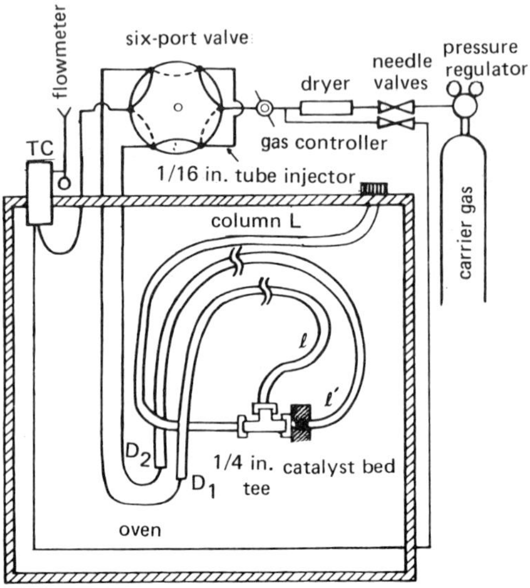

Fig. 6.6 Gas lines and important connections of the experimental arrangement used to feed the reactant(s) onto the catalyst by diffusion (10). TC is a thermal conductivity detector, but other detectors can be used as well. For hydrogenation reactions the carrier gas is H_2.

The diffusion column L is empty of any solid material, whereas sections l' and l are filled with chromatographic material except for a short length (5–10 mm) near the junction with column L which contains the catalyst. This part is heated, if necessary, with a separate heating element at a temperature different from that of the chromatograph's oven, and is recorded with a separate thermocouple.

If only section l is filled with chromatographic material, the pressure drop along it must be small, so that the pressure change in tube L at each flow reversal is small. However, this pressure change lasts only for a short time equal to the duration of the backward flow, t'. In cases of large pressure drops along l,

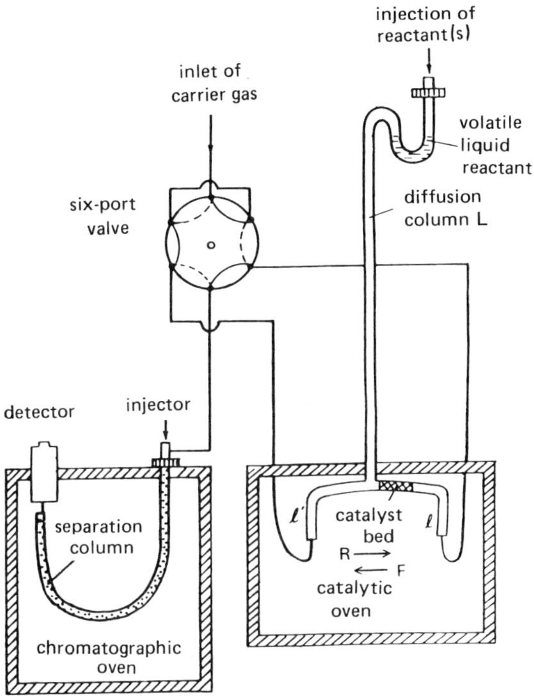

Fig. 6.7 Outline of an experimental setup for catalytic studies using a diffusion feed and a separation column other than the sampling column l' + l (6).

the column section l' should also be filled with the same material and be of the same length as l to minimize pressure changes at the exit of L.

The arrangement of Fig. 6.7 is preferable, since it has the following noticeable features: 1. The flow reversals are "confined" in the sampling column l' + l placed inside the catalytic oven, whereas in the separation column found in the chromatographic oven the carrier gas flows always in the same direction, from injector toward the detector. As a result of this, the pressure changes at the exit of column L are negligible, since the catalyst bed is of very short length and the pressure drop along it very small; 2. reference substances for identification purposes can be chromatographed in the separation column by introducing them through the injector of the chromatographic oven; 3. the catalyst can be heated at a quite different temperature than that of the separation column; and 4. a restrictor like H in Fig 6.5 is no longer necessary.

The conditioning of the catalyst is carried out in situ at an appropriate temperature for 15–24 h, under continuous carrier gas flow in both directions F and R, and with some preliminary injections of reactant to establish a constant catalytic activity.

Following the above conditioning, and after the chromatographic signal had decayed to a negligible height, a new volume of reactant at atmospheric pressure is injected into the column L. After 10-20 min, a continuous concentration-time curve decreasing slowly is established in the recorder owing to both the reactant and the products. This can be taken as the steady-state condition for the catalyst. Then, the carrier gas flow is reversed from the F to the R direction by turning the valve from one position (solid lines) to the other (dashed lines). After about 30 s of backward flow, the carrier gas is turned again to the original F direction. If the time, t', elapsing between two successive reversals of the flow (one from F to R direction and one back to F direction) is less than the retention time on columns l' and l of the substances giving rise to the chromatographic signal, a symmetrical sample peak for each substance follows the restoration of the gas flow to its original direction, as shown in Fig. 6.4. The above procedure of the flow reversals is repeated many times at each temperature, giving rise to a series of sample peaks which may correspond to differential partial pressures of reactant(s) over the catalyst and different fractional conversions of reactants to products.

The diffusion technique described above offers the following advantages over traditional techniques used in heterogeneous catalysis, outlined below.

1. It is a very simple setup and instead of using flowmeters, mixing chambers, saturators, and similar devices to introduce the reaction mixtures into the catalyst bed, a column L empty of any solid material is simply filled with the one reactant using a gas-tight syringe (or with a mixture of two reactants), and this reactant is allowed to diffuse slowly onto the catalyst retained within a short section near the junction of columns L and l' + l.
2. The diffusion feed ensures the presence of reactants over the catalyst bed for a long time period, resembling that of continuous flow, with the concentration (and hence partial pressure) of reactant(s) either changing continuously with time, according to the diffusion laws, or remaining constant.
3. Reversing the flow of the carrier gas by means of the valve for a short time period t', causes a sampling of all substances present at $x = l'$. The chromatographic material filling l separates these substances, revealing their relative concentrations after passing through the catalyst bed in the form of extra peaks (sample peaks). From these sample peaks, the fractional conversion of reactants to products can be calculated many times during a single diffusion feed. From these conversions, covering a wide range, the rate constant and the order of the reaction can be found for an extending range of partial pressures <u>in a single experiment</u> lasting only a few hours.
4. The catalytic bed can be treated either as a differential reactor or in an integral mode.
5. It is apparent from Fig. 6.4 that the method uses "pulses" of reactants created by the flow reversals, but these pulses are superimposed on an initial concentration line of reactant and product(s), constituting a continuous flow of substances over the catalyst bed. Thus it has all the advantages of both a pulse and a continuous feeding of reactant technique, and because this feeding is due to slow gas diffusion, steady-state

conditions over the catalyst are soon and easily established. In other words, the method is a pulse technique under steady-state conditions. This would lead to kinetic information not masked by adsorption effects of reactants and products on the catalyst surface.

III. REACTION KINETICS WITH THE PULSE TECHNIQUE

A. Theory

If we have the experimental arrangement described by Figs. 6.1 and 6.5, i.e., the reactant is injected as a pulse at a middle position of column $l' + l$ and retained there because of its high adsorption equilibrium constant on the catalyst, the rate of the reaction is written as $r(t_0)\,\delta(x - l')$, the delta function representing the x distribution of the reaction rate. Then $r(t_0)$ is only a function of time expressed in mol cm^{-2} s^{-1}; i.e., per unit cross-section a of the void space in the column. Identifying $r(t_0)\,\delta(x - l')$ with the last term in Eq. (3-1), $vc(l',t_0)\delta(x - l')$, we obtain

$$c(l',t_0) = \frac{r(t_0)}{v} = \frac{r(t_0)a}{\dot{V}} \tag{6-1}$$

This is a product concentration due to the chemical reaction. The gaseous concentration of the reactant is practically zero, since it has a high adsorption equilibrium constant.

An example of sample peaks due to a chemical reaction obtained with the pulse technique has been given in Fig. 6.2. This should be compared with the theoretical elution curves of Fig. 3.5.

If an assumption concerning the mechanistic model of the reaction can be made, the reaction rate, $r(t_0)$, can be expressed as an analytic function of time and this can be substituted directly in Eq. (6.1). The simplest possible case of mechanistic models is

$$A + S \underset{K_A}{\overset{\text{fast}}{\rightleftharpoons}} A - S \xrightarrow{k} D - S \underset{K_D}{\overset{\text{fast}}{\rightleftharpoons}} D + S \qquad (6\text{-}2)$$

i.e., the reactant, A, is rapidly adsorbed on the active centers, S, giving the adsorbed reactant, A — S. This is then decomposed by a first-order step with a rate constant, k, to give the adsorbed product, D — S, which rapidly equilibrates with the gaseous product, D. The K_A and K_D are adsorption equilibrium constants of A and D, respectively, and if $K_A >> K_D$, the rate of formation of D in mol $cm^{-2}s^{-1}$ is given by

$$r(t_0) = \frac{kmg}{a} \cdot \exp(-kt_0) \qquad (6\text{-}3)$$

where m is the amount of A injected (mol) and g its fraction on <u>reactive</u> sites of the surface. The right-hand side of Eq. (6-3) can now be substituted for $r(t_0)$ in Eq. (6-1) and the result

$$c(l',t_0) = \frac{kmg}{\dot{V}} \cdot \exp(-kt_0) \qquad (6\text{-}4)$$

is used in Eq. (3-21) to express c_1 and c_2 analytically:

$$c = \frac{kmg}{\dot{V}} \Big\{ \exp[-k(t_{tot} + \tau)] \cdot u(\tau)$$

$$+ \exp[-k(t_{tot} - \tau)] \cdot [u(\tau) - u(\tau - t'_R)] \Big\} \qquad (6\text{-}5)$$

Here c is the concentration of the product, D, at the detector, $t_{tot} = t_0 + t'$, $\tau = t - t_R$ [Eq. (3-20], t_R and t'_R being the retention times of D on the column sections l and l', respectively.

The area under the R sample peaks, shown shaded in Fig. 6.2, is obtained by applying Eq. (3-23). In place of $C'(l,p_0)$ we put $C(l',p_0)$, which is the Laplace transform of Eq. (6-4). The final result is obtained by reversing the transformation with respect to p_0:

$$f = mg[\exp(kt_R) - 1] \cdot \exp(-kt_0) \qquad (6\text{-}6)$$

For the F-peaks, t_R' must be substituted for t_R. The physical meaning of this equation becomes obvious if it is written as $f = mg \exp[-k(t_0 - t_R)] - mg \exp(-kt_0)$. It means that when the direction of the gas flow is reversed at time t_0, the product formed between the times $t_0 - t_R$ and t_0 is exhibited as a sample peak. It follows that f is proportional to t_R (or t_R' for the F-peaks), and thus a short column length, l or l', produces small narrow peaks, such as peak 2 in Fig. 6.2. According to Eq. (6-6), a plot of ln f versus t_0 gives k from the slope of the straight line obtained, if the reaction conforms to the model (6-2).

When plotting Eq. (6-6) to find k from the slope, a relative value for the expression $mg[\exp(kt_R) - 1]$ having the same units as f is obtained. If the response of the detecting system for the product measured can be estimated, e.g., by injecting known amounts of the pure product into an empty column under the same experimental conditions, then the absolute value (mol) of the above expression can be found. From the known values of m, t_R, and k, the fraction of active catalytic surface, g, can be computed. The deviation of its mean value from unity explains why the % converstion to products deviates considerably from 100%, in spite of the fact that the reactant is never eluted from the catalytic column: It is due to the fact that % conversion is simply equal to 100g. Thus dividing f by t_R, an average reaction rate is obtained and then

$$\text{Conversion} = \frac{1}{m}\int_0^{\infty} \frac{f}{t_R}\, dt_{tot} = \frac{g[\exp(kt_R) - 1]}{kt_R} \simeq g \qquad (6\text{-}7)$$

The last approximation is based on the relation $\exp(x) \simeq 1 + x$ for small x.

That the relatively small value of g is actually due to the small fraction of the total surface that is catalytically active, and not to irreversible adsorption of the product on the solid catalyst, is proved by the fact that two column sections (l' and l) containing very different amounts of catalyst give virtually the same value of g, as shown in the following paragraphs.

B. Dehydration of Alcohols

An example of plotting Eq. (6-6) is given in Fig. 6.8. It refers to the dehydration of propan-2-ol over 13X molecular sieve to give diisopropyl ether (2).

The rate constant k, calculated from the plots of R- and F-peaks in Fig. 6.8, is $(2.5 \pm 0.1) \cdot 10^{-4}$ and $(2.6 \pm 0.1) \cdot 10^{-4}\ s^{-1}$, respectively. Rate constants at other temperatures and other alcohols over 13X molecular sieve and γ-aluminium oxide have been calculated and reported (3). They all obey Eq. (6-6). A t-test of significance performed on the coefficients of regression of ln f on t_0 shows that these are significant at a level better than 1%, indicating that the probability for the corresponding t-value of being exceeded is <1%. This is a measure of the goodness of fit of the experimental data by linear plots like those of Fig. 6.8.

In Fig. 6.9 Arrhenius plots for the dehydration of one alcohol are shown. From such plots the activation parameters given in Table 6.1 have been calculated (3). Note that these are true activation parameters, not involving heats of adsorption, since they are derived from rate constants pertaining to the surface step of the reactions. It is evident from Table 6.1 that activation parameters determined from R-peaks coincide, within the limits of experimental error, with those found from F-peaks. This is important, because the lengths l and l' of the chromatographic column responsible for the R- and F-peaks, respectively, differ by a factor of 14 or more. This coincidence indicates that secondary reactions of the detected product or irreversible adsorption of it are negligible.

The activation energies found are consistent with some literature values. Thus, Gentry and Rudham (12) have determined an

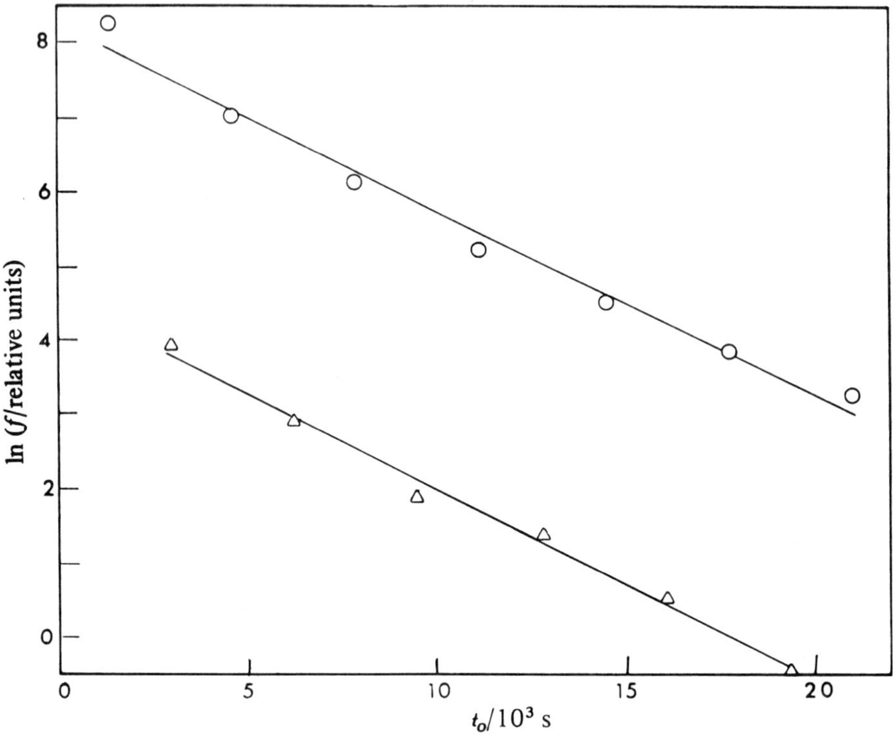

Fig. 6.8 Plots of Eq. (6-6) for the dehydration of propan-2-ol (0.5 mm^3) to di-isopropyl ether, at 452°K, over 13X molecular sieve (80-100 mesh) activated at 673°K for 12 h. The lengths of the glass column were l' = 7 and l = 100 cm, and the carrier gas nitrogen ($\dot{V}$ = 0.446 cm^3s^{-1}). o, R-peaks; Δ, F-peaks (2).

activation energy of 110 kJ mol^{-1} for the dehydration of propan-2-ol to di-isopropyl ether over a 13X molecular sieve surface similar to that of Ref. (3).

The amount of the reactant alcohol and the volume flowrate of the carrier gas seem to have no significant effect on the rate constants and the activation energies (3).

The fraction of active catalytic surface, g, for the system propan-2-ol/13X molecular sieve has been calculated, as explained

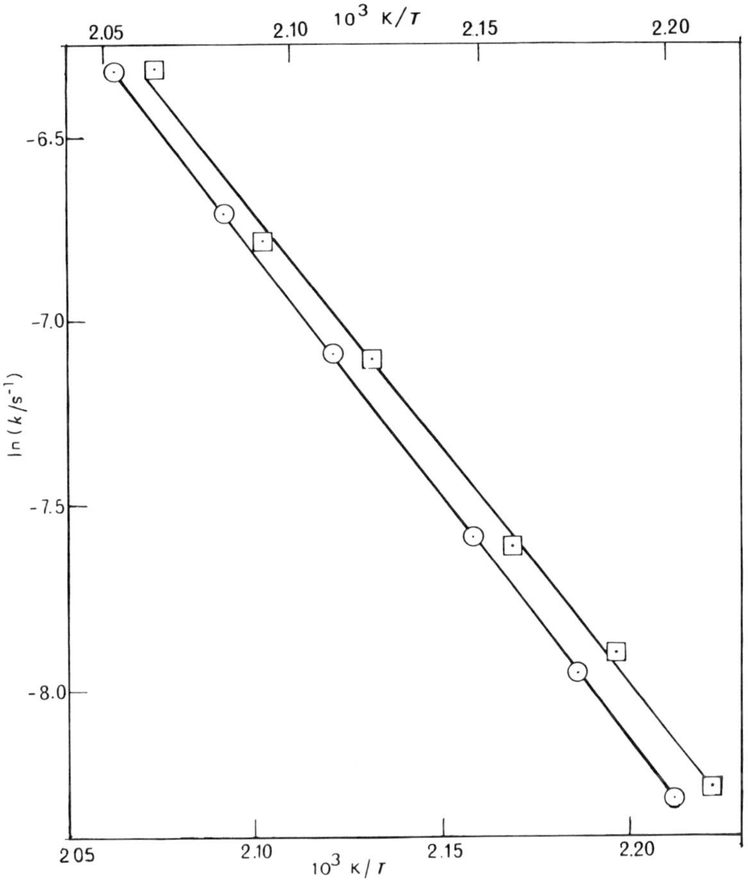

Fig. 6.9 Arrhenius plots for the dehydration of propan-2-ol to di-isopropyl ether over 13X molecular sieve. The experimental conditions are those of Fig. 6.8. o, Rate constants determined from R-peaks (lower abscissa); □, rate constants from F-peaks (upper abscissa) (2).

in paragraph A of this section, and the results at five temperatures between 452 and 485°K show that the g value does not seem to change significantly with temperature or from R- to F-peaks (3). Its mean value is around 0.25, which deviates considerably

Table 6.1 Activation Energies (E_a) and Frequency Factors (A) for the Dehydration of Three Alcohols over 13X Molecular Sieve and γ-Aluminum Oxide

				E_a (kJ mol^{-1})		ln (A/s^{-1})	
Alcohol/catalyst	Main product	l' (cm)	l (cm)	R-peaks	F-peaks	R-peaks	F-peaks
Propan-1-ol/13X	Propene	3.0	108	96 ± 4	98 ± 4	16 ± 1	16.1 ± 0.9
Propan-2-ol/13X	Di-isopropyl ether	7.0	100	110.6 ± 0.3	107 ± 3	21.1 ± 0.1	20.1 ± 0.7
Butan-1-ol/13X	But-1-ene	1.3	45	118 ± 8	115 ± 8	19 ± 2	19 ± 2
Propan-1-ol/Al_2O_3	Propene	1.5	45	78 ± 5	82 ± 6	10 ± 1	10 ± 1
Propan-2-ol/A_2O_3	Propene	1.5	45	78 ± 6	81 ± 7	12 ± 1	13 ± 2

Source: Ref. (3).

from unity, explaining the experimental deviation of the conversion to products from 100%.

C. Deamination of Primary Amines

The same experimental setup (Fig. 6-5) was used to study the kinetics of deamination of some primary amines; i.e., 1. of 1-aminopropane and 2-aminopropane to propene over 13X molecular sieve (4), and 2. of aminocyclohexane to cyclohexene over 13X molecular sieve (4) and over γ-aluminium oxide (2).

The reactions over 13X molecular sieve conform to the same mechanistic model as that applied to the dehydration of alcohols (Eq. [6-2]). Eq. (6-6) was used again to determine rate constants at various temperatures. From these the activation parameters collected in Table 6.2 were computed.

The differences between activation parameters determined from R- and from F-peaks lie again within the limits of experimental error, showing that secondary reactions of the detected product or irreversible adsorption of it are negligible. This is because the two lengths, l and l', of the column were very different (by a factor of 16-36).

No comparison of these results with literature values can be made, since to the best of our knowledge, no deaminations over zeolites have been studied previously.

An increasing conversion of aminocyclohexane to cyclohexene with increasing working temperature, as judged from g, is observed here. This again is not due to irreversible adsorption of the product on the solid catalyst, as the g values calculated from R- and F-peaks were not significantly different.

The reaction of aminocyclohexane over γ-aluminum oxide follows a different mechanistic model, since the time dependence of the peak height at $\tau = 0$ is that shown in Fig. 6.10. This suggests consecutive first-order reactions:

Table 6.2 Activation Energies (E_a) and Entropies ($\Delta S^{\neq}$) for the Deamination of Three Amines over 13X Molecular Sieve

			E_a (kJ mol^{-1})		$-\Delta S^{\neq}$ (JK^{-1} mol^{-1})	
Amine	l' (cm)	l (cm)	R-peaks	F-peaks	R-peaks	F-peaks
1-Aminopropane	3	108	71 ± 2	66 ± 9	212 ± 3	218 ± 17
2-Aminopropane	0.5	8	144 ± 3	138 ± 3	84 ± 6	94 ± 6
Aminocyclohexane	1.3	46	162 ± 11	156 ± 5	51 ± 17	59 ± 8

Source: Ref. (4).

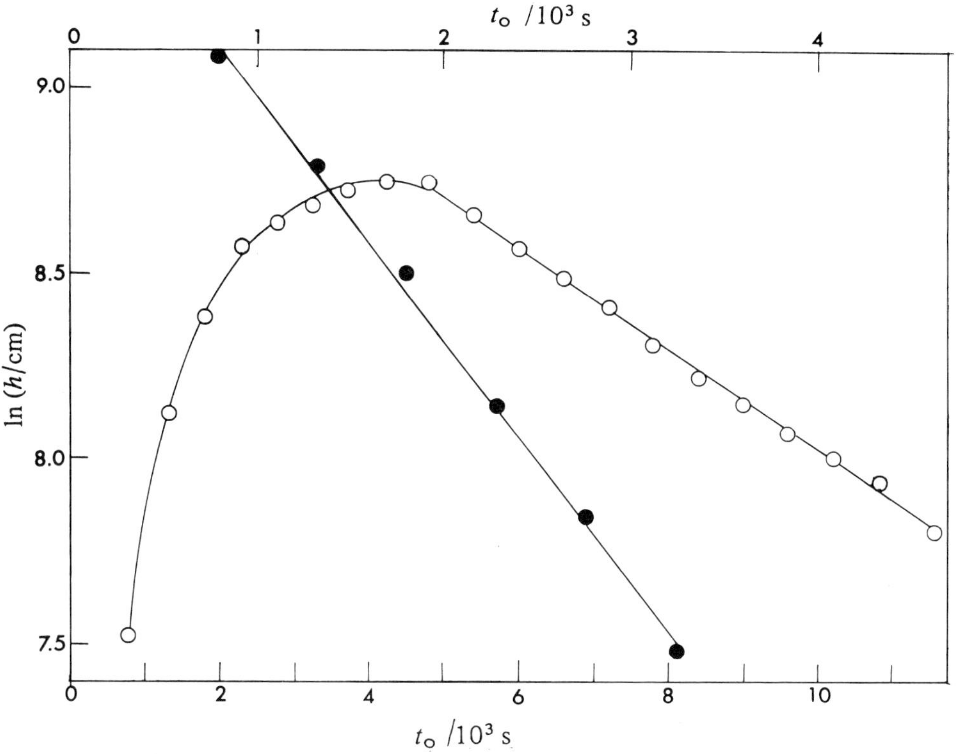

Fig. 6.10 Time-dependence of the ln h values for R-peaks at $\tau = 0$, for the deamination of aminocyclohexane (1 mm^3) to cyclohexene at 525°K, over γ-aluminum oxide (60-70 mesh, Ho-415 Houdry-Hüls) activated at 673°K for 2 h. Both lengths l' and l of the glass column (i.d. 4 mm) were 40 cm, and the carrier gas nitrogen ($\dot{V} = 0.30\ cm^3\ s^{-1}$). o, (lower abscissa) experimental points h; •, (upper abscissa) values of h' obtained by subtracting h from the corresponding extrapolated values Q exp $(-k_1 t_0)$ of the last linear part back to the times of the ascending experimental curve (2).

$$A + S \underset{K_A}{\overset{\text{fast}}{\rightleftharpoons}} A - S \xrightarrow{k_1} B - S \xrightarrow{k_2}$$

$$D - S \underset{K_D}{\overset{\text{fast}}{\rightleftharpoons}} D + S \tag{6-8}$$

with a surface intermediate B − S.

The rate of production of the final product D is now proportional to the concentration of the adsorbed B − S and this is given by the classic equation for an intermediate in consecutive reactions. Thus

$$r(t_0) = \frac{k_1 k_2 mg}{a(k_2 - k_1)} [\exp(-k_1 t_0) - \exp(-k_2 t_0)] \tag{6-9}$$

As before, the right-hand side of this expression is substituted for $r(t_0)$ in Eq. (6-1), and the result

$$c(l', t_0) = Q[\exp(-k_1 t_0) - \exp(-k_2 t_0)] \tag{6-10}$$

where

$$Q = \frac{k_1 k_2 mg}{\dot{V}(k_2 - k_1)} \tag{6-11}$$

is used in Eq. (3-21) to write c_1 and c_2 explicitly:

$$\begin{aligned} c = {} & Q\{\exp[-k_1(t_{tot} + \tau)] - \exp[-k_2(t_{tot} + \tau)]\} \cdot u(\tau) \\ & + Q\{\exp[-k_1(t_{tot} - \tau)] - \exp[-k_2(t_{tot} - \tau)]\} \\ & \cdot [u(\tau) - u(\tau - t'_R)] \end{aligned} \tag{6-12}$$

where $t_{tot} = t_0 + t'$. To determine the rate constants k_1 and k_2, we can employ Eq. (3-24) which, by using Eq. (6-10) in place of c_2, gives

$$h = Q[\exp(-k_1 t_{tot}) - \exp(-k_2 t_{tot})] \tag{6-13}$$

in agreement with the general appearance of the experimental curve of Fig. 6.10. From the slope of the last linear part (after the induction period) the smaller rate constant, say k_1, was found equal to $(1.36 \pm 0.02) \cdot 10^{-4}\ s^{-1}$. Then, the term Q exp $(-k_2 t_{tot})$ is calculated from the difference $h' = Q \exp(-k_1 t_{tot}) - h$, i.e., by extrapolating back the last linear part and subtracting from it the experimental values of h. If now ln h' is plotted versus t_{tot}, k_2 is found from the slope of this new linear plot, as $(6.5 \pm 0.2) \cdot 10^{-4}\ s^{-1}$. Analogous results were found from the F-peaks (2).

The absolute value of Q/mol cm^{-3} in Eq. (6-11) can be determined by the relation

$$Q_{abs}/\text{mol cm}^{-3} = \frac{(Q_{rel}/\text{cm})}{\dot{V}S} \qquad (6\text{-}14)$$

where S/cm s mol^{-1} is the response of the detecting system. From Q_{abs} and Eq. (6-11), g was found equal to 0.116, showing again that the active catalytic surface is only a small fraction of the total surface. This possibility has been pointed out many times in the literature. A relation analogous to Eq. (6-7), but based on the peak height, h, is

$$\text{Conversion} = \frac{1}{m} \int_0^\infty h dV = \frac{1}{m} \dot{V} \int_0^\infty h dt_{tot} = g \qquad (6\text{-}15)$$

where Eq. (6-13) was used for h.

D. Catalytic Cracking of Cumene

The catalytic cracking of cumene over zeolite catalysts has been the subject of various investigations concerning the active sites on the catalyst surface, identification and yields of primary and secondary products, and the mechanism of the reactions. The latter has been based on kinetic studies performed with three common reactor types: pulse microcatalytic, differential, and integral. Best and Wojciechowski (13) have tabulated the results obtained with these three reactor types by various authors. Prater and Lago (14) were among the first who proposed a mechanism for the dealkylation of cumene to benzene and propene. The mechanism was expanded by Campbell and Wojciechowski (15) to the so-called delta mechanism shown in Fig. 6.11. Based on this mechanism, rate expressions, in terms of fractional conversions and various model parameters, were derived (5) for the cases in which the rate-controlling step is the bond-breaking step (15), the desorption of reaction producers (16), or the adsorption of cumene (16). These expressions were extensively used by Wojciechowski and his coworkers in various kinetic

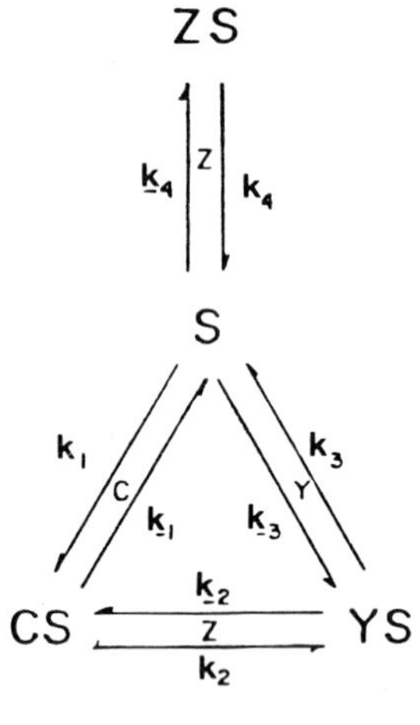

Fig. 6.11 The delta mechanism of Campbell and Wojciechowski (15). C, Y, and Z represent cumene, propene, and benzene, respectively, S the catalyst active sites, and CS, YS, and ZS the corresponding chemisorbed species. The various k's are rate constants for the chemical reaction and the adsorption-desorption processes shown (5).

studies of the dealkylation and other reactions occurring in cumene cracking over LaY and HY zeolites (13,16-24).

The rate equations mentioned above, however, are complicated expressions involving many parameters, defined in terms of other parameters (mainly adsorption equilibrium constants). The fit of these equations to the experimental data, leading to determination of model parameters or activation energies is a rather difficult task, solved only with the aid of a powerful computer. By contrast, the reversed-flow gas chromatography technique is a much simpler differential method to study the kinetics in heterogeneous catalysis. Its equations are very simple, containing only the rate constants of the various steps as parameters. Their fit to the experimental data is straightforward, leading to the determination of true rate constants and activation energies, which are not apparent values involving heats of adsorption. In the present paragraph, the method is applied to catalytic cracking. The dealkylation of cumene over LaY and HY zeolites was chosen, giving propene and benzene as the main products (5). The experimental setup and the procedure followed is the same as in the previous reactions. The catalyst was contained in two gas chromatographic columns, l' and l, connected in series with an injector between them. In this work, the lengths l' + l were 14.3 cm (i.d. 4 mm) + 119.8 cm (i.d. 1.6 mm) containing 1.41 + 1.74 g LaY catalyst, and 2 + 43.6 cm (both of i.d. 4 mm) containing 0.15 + 3.77 g base LZ−Y82.

The conditioning of the columns was carried out in situ under carrier gas flow (0.33 and 0.65 $cm^3\ s^{-1}$ for the first and the second column, respectively). The starting temperature was 373°K and this was increased at a rate of 11°K/min up to 683°K, where the columns were kept for 20 h. During this period, the base LZ−Y82 lost ammonia and was transformed to the HY zeolite.

Following each activation, the columns were cooled to the working temperature, and after some preliminary reactant injections, 2 — 5 mm^3 of liquid cumene were introduced with a microsyringe through the injector placed between the two columns. The carrier gas was flowing, with the same volumetric rates mentioned above, in the direction F; i.e., entering the short column, l', and leaving the long column, l, toward the detector. The time interval between any two successive reversals of the gas flow was always greater then the sum $t'_R + t_R$ of the retention times of the products on the column sections l' + l.

Identification of the products was made by injecting reference substances, and also with the help of a time of flight mass spectrometer. Since the reactant cumene (C) is introduced as a pulse at a middle position of the whole column l' + l, and the chromatographic process on C is repeatedly changing direction (forward and backward), the same approximation as before is adopted; i.e., that the distance x distribution along the column of the reaction rate $r(x, t_0)$, is described by a delta function $\delta(x - l')$. Thus we can write for the production rate of a substance $r(t_0)$ $\delta(x - l')$, where $r(t_0)$ is only a function of time t_0 from the injection of C, and is expressed in mol · cm^{-2} s^{-1}, as before.

The general equation giving the area f under the R-sample peaks of a product is given by Eq. (3-23). This combined with Eq. (6-1) gives

$$\mathscr{L}_{t_0} f = \frac{aR(p_0)}{p_0} [1 - \exp(-p_0 t_R)] \tag{6-16}$$

where $R(p_0)$ is the transformed rate of production, $r(t_0)$, of a substance, and t_R the retention time of the product on column l. For the F-peaks, obtained with the carrier gas flowing in the opposite direction, the same equation applies with the retention time, t'_R, on column l' substituted for t_R.

To apply Eq. (6-16) in the dealkylation reaction of cumene over LaY and HY zeolites, we must determine the rate of production of these products, assuming a certain mechanistic model for the reaction. Then substituting the Laplace transformed rate, $R(p_0)$, in Eq. (6-16) and taking the p_0 inverse transform, the area f of the sample peaks will be found as an analytic function of the time t_0. This function will then be fitted to the experimentally determined peak areas to calculate the rate constants of the slow rate-controlling step(s).

If we adopt the delta mechanism of Campbell and Wojciechowski (Fig. 6.11) as the working model, and neglect the back reaction (k_{-2} branch), since the method is a differential one, we write for the rate of disappearance of C:

$$-\frac{dc_C}{dt_0} = k_1 c_S c_C - k_{-1} c_{CS} \tag{6-17}$$

the rate of change of CS:

$$\frac{dc_{CS}}{dt_0} = k_1 c_S c_C - k_{-1} c_{CS} - k_2 c_{CS} \tag{6-18}$$

the rate of change of YS:

$$\frac{dc_{YS}}{dt_0} = k_2 c_{CS} + k_{-3} c_S c_Y - k_3 c_{YS} \tag{6-19}$$

and the rate of production of Y:

$$\frac{dc_Y}{dt_0} = k_3 c_{YS} - k_{-3} c_S c_Y \tag{6-20}$$

where the various c are molar concentrations of the species shown as subscripts.

Since the amount of catalyst is relatively large, whereas the amount of the reactant very small, the concentration c_S can be regarded as a constant and combined with k_1 in Eqs. (6-17) and (6-18) or with k_{-3} in Eqs. (6-19) and (6-20) to give a new constant:

$$k_1' = k_1 c_S \qquad k_{-3}' = k_{-3} c_S \tag{6-21}$$

The system of differential equations (6-17) to (6-20) can be transformed to a system of algebraic equations by Laplace transformation with respect to t_0, using the initial conditions:

$$c_C(0) = c_0$$

$$c_{CS}(0) = c_{YS}(0) = c_Y(0) = 0 \tag{6-22}$$

Solving this new system of equations for C_Y (the Laplace transformed function of c_Y), we obtain

$$C_Y = \frac{k_2 c_0}{A\ B\ (p_0 + k_1')\ (p_0 + k_3)} \tag{6-23}$$

where

$$A = \frac{p_0 + k_{-1} + k_2}{k_1'} - \frac{k_{-1}}{p_0 + k_1'} \tag{6-24}$$

and

$$B = \frac{p_0 + k'_{-3}}{k_3} - \frac{k'_{-3}}{p_0 + k_3} \tag{6-25}$$

The rate of formation of Y is simply $r = V(dc_Y/dt_0)/a$, where V is the volume of the gas phase in the column. The Laplace transform of the rate, under the initial condition (6-22), is $R = Vp_0C_Y/a$. This substituted in Eq. (6-16) gives

$$\mathscr{L}\ t_0{}^{f}y = V\ C_Y[1 - \exp\ (-\ p_0 t_R)] \tag{6-26}$$

The inverse Laplace transform of this expression, with Eq. (6-23) substituted for C_Y, gives the area f_Y under the R-peaks (or under the F-peaks with t_R' in place of t_R), as an analytic function of the time t_0, for any step of Fig. 6.11 being rate controlling, or even for all steps influencing the rate of appearance of Y.

Clearly, Eq. (6-26) gives also the area under the R- and F-peaks, if these consist of the other product, Z. In this case, C_Y in Eq. (6-26) must be substituted by C_Z, and this is given by an expression analogous to (6-23) with k_4 and $k'_{-4}(= k_{-4}c_S)$ substituted for k_3 and k'_{-3}, respectively.

Two limiting cases of Eqs. (6-23) and (6-26) are interesting. If only one of the steps of the mechanism of Fig. 6.11 is slow and rate controlling, whereas the others are fast, Eq. (6-23) is greatly simplified by omitting p_0 compared with the rate constants of the fast steps. This simplified equation is then substituted for C_Y in Eq. (6-26), giving an expression whose inverse Laplace transform is easily found. Let us take, for example, the k_2 step (the dealkylation reaction) as rate controlling, assuming the adsorption of C (k_1 step) and the desorption of YS (k_3 step) fast. Omitting p_0 as compared with k_1' and k_3, Eq. (6-23) becomes

$$C_Y = \frac{k_2 c_0}{AB\ k_1' k_3} \tag{6-27}$$

with A and B now given by

$$A = \frac{p_0 + k_2}{k_1'} \qquad B = \frac{p_0}{k_3}$$

This substituted for C_Y in Eq. (6-26) gives

$$\mathscr{L}\ t_0 f_Y = \frac{V\ k_2 c_0}{p_0(p_0 + k_2)}\ [1 - \exp(-p_0 t_R)] \tag{6-28}$$

and the inverse transform of this expression (for $t_0 > t_R$) is

$$f_Y = Vc_0[\exp(k_2 t_R) - 1] \cdot \exp(-k_2 t_0) \tag{6-29}$$

Had adsorption of C (k_1 step) been taken as rate controlling, p_0 would be omitted compared with k_2 and k_3 in Eq. (6-23), and the final result would again be Eq. (6-29), where in place of k_2 an apparent rate constant k_a would appear

$$k_a = \frac{k_1'}{1 + k_{-1}/k_2} \tag{6-30}$$

Finally, if desorption of Y is rate controlling (k_3 step), the following equation would be obtained instead of (6-29):

$$f_Y = \frac{Vk_3c_0}{k_s}[\exp(k_st_R) - 1] \cdot \exp(-k_st_0) \tag{6-31}$$

where $k_s = k_3 + k'_{-3}$. Eqs. (6-29) and (6-31) show that if one step of Fig. 6.11 is rate controlling, a linear plot of ln f_y against t_0 is predicted. From the slope of this plot a rate constant can be determined, but it is impossible to ascertain whether this is k_2 or k_a or k_s, since the final equation giving f_y as a function of t_0 has exactly the same form in all three cases. This was also the case with the rate expressions derived by Best and Wojciechowski (16) using the same mechanism.

A second limiting case of Eqs. (6-23) and (6-26) is that the overall reaction rate is controlled by <u>two</u> steps in the mechanism of Fig. 6.11; i.e., only one of the three steps (k_1, k_2, or k_3) is fast compared with the other two. This possibility has never been explored before owing obviously to the very complicated rate equations which would result. This is not the case, however, with the technique employed here. Let us assume, for example, that only adsorption of C (k_1 step) is fast, whereas both other steps (k_2 and k_3) are slow. Then, omitting p_0 compared with k_1' only, Eq. (6-23) becomes

$$C_Y = \frac{k_2k_3c_0}{p_0(p_0 + k_2)(p_0 + k_3 + k'_{-3})} \tag{6-32}$$

This is substituted for C_Y in Eq. (6-26) as before, and the inverse Laplace transformation is found for $t_0 > t_R$:

$$f_Y = \frac{Vk_2k_3c_0}{k_2 - k_s}\left[\frac{\exp(k_st_R) - 1}{k_s} \cdot \exp(-k_st_0) - \frac{\exp(k_2t_R) - 1}{k_2} \cdot \exp(-k_2t_0)\right] \quad (6\text{-}33)$$

where $k_s = k_3 + k'_{-3}$. The factors before exp $(-k_st_0)$ and exp $(-k_2t_0)$ inside the brackets [ ] are both approximately equal to t_R if k_s and k_2 are sufficiently small.

The two other possibilities of having two slow steps are when only the k_2 step or only the k_3 step is fast, the remaining two steps in each case being slow. Both these cases lead to the same integrated rate equation as Eq. (6-33), with different rate constants. Therefore, the mere fitting of these equations to the experimental data does not distinguish between the three above possibilities. It does distinguish, however, a mechanism with two slow steps from another with one such step, since the behavior of Eq. (6-33) is quite different from that of Eq. (6-29) or (6-31). A product whose production rate follows Eq. (6-33) exhibits an induction period, its amount increasing initially with time, passing through a maximum and then decreasing almost exponentially with a rare coefficient equal to the smaller rate constant.

With LaY as catalyst both products of the dealkylation reaction, Y and Z (propene and benzene, respectively), appear on the reversed-flow chromatogram forming separate sample peaks with different retention times. The peaks of propene decrease with time from the outset, according to a simple exponential law, as Fig. 6.12 shows. Obviously, the formation of propene follows Eq. (6-29) or Eq. (6-31), and, therefore, a limiting case of the delta mechanism with only one step being rate controlling. From the slopes of plots such as that of Fig. 6.12, the fate constants of Table 6.3 were calculated by standard least-squares procedures.

The sample peaks of benzene have significant height only in the R direction. As shown previously (2), this is due to the large

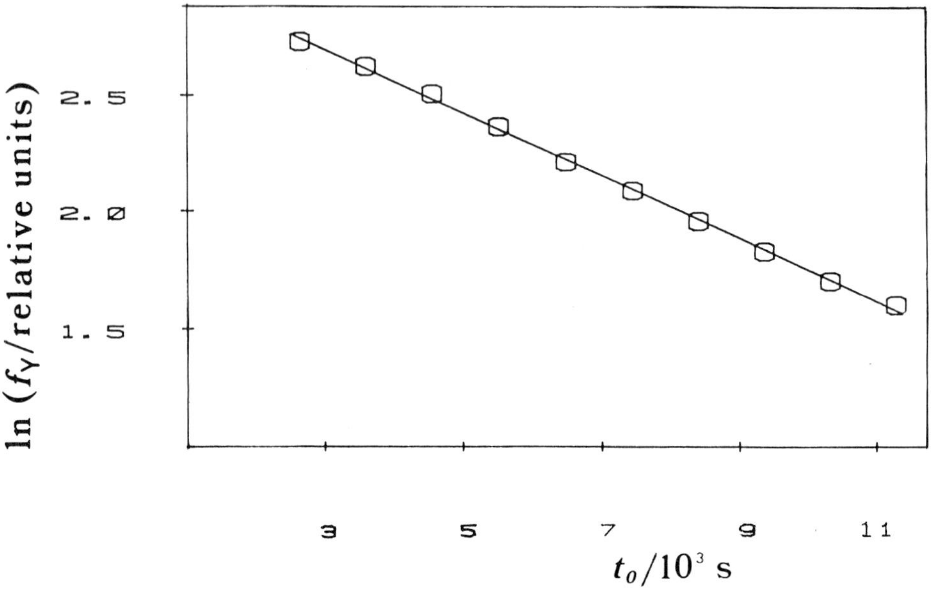

Fig. 6.12 Plot of Eq. (6-29) or (6-31) for the cracking of cumene over LaY catalyst, at 539°K with a carrier gas flowrate of 0.33 $cm^3\ s^{-1}$. The f_Y values are the areas under the F-peaks of propene (5).

Table 6.3 Kinetic Parameters for Propene Formation During Dealkylation of Cumene Over LaY Catalyst

T(K)	$10^4 k(s^{-1})$	E_a (kJ mol^{-1})	ln (A/s^{-1})
525	0.52 ± 0.01	135 ± 4	21 ± 1
539	1.30 ± 0.02		
546	1.75 ± 0.01		
551	2.37 ± 0.04		
561	3.8 ± 0.1		

Source: Ref. (5).

t_R value of benzene on column 1 as compared to the very small t'_R value on column 1'. The area f_Z under these peaks initially increases with time until a maximum value is reached, and then decreases. This behavior is predicted by Eq. (6-33), but not by Eq. (6-29). Thus the formation of benzene involves two slow steps in the delta mechanism. One of these steps must be the slow desorption of benzene with a rate constant $k_S = k_4 + k'_{-4}$, since adsorption of cumene and the bond-breaking act are steps common to both products, propene and benzene. Had both of these steps been slow, the formation of propene would also follow Eq. (6-33) and not Eq. (6-29).

There is some indication that the k values of Table 6.3 correspond to k_2 because the activation energy of 135 kJ mol^{-1}, calculated from the Arrhenius plot of Fig. 6.13, is rather high to be connected with k_a or k_S, which represent adsorption–desorption processes. The plots of Fig. 6.13 indicate that both LaY and HY used are diffusion-free catalysts.

An implication of k_2 being a true rate constant is that the value of the frequency factor found is due to the entropy of activation, which is calculated as -84 JK^{-1} mol^{-1}. This negative value indicates that the transition state $CS^{\neq}$ on the catalyst surface is more localized or generally has less freedom on the surface than the adsorbed species CS. Conventional rate equations like those of Wojcienchowski et al. (15,16) lead to the determination of $k_2 c_S$ and Arrhenius plots based on this combination cannot give the entropy of activation.

The reversed-flow chromatograms with HY as catalyst show only the peaks of benzene, whereas those of propene are delayed, obviously because of the formation of carbonium ions on the catalytic surface (17). These ions are probably very slowly decomposed because of the high concentration of Brönsted sites on this catalyst. From the F- and R-peaks of benzene, which follow Eq. (6-29) or (6-31), rate constants in both directions were determined and are collected in Table 6.4. It is noteworthy that the % difference between the two k's at the same temperature is not at all significant, in spite of the fact that the F- and the R-peaks are due to such different column lengths, l' and l (2 and 43.6 cm, respectively), containing so much different amounts of catalyst (0.15 and 3.77 g, respectively). This has also been noted earlier (3,4) for other reactions.

Using the mean k value at each temperature, the Arrhenius plot for HY was constructed and is given in Fig. 6.13. The activation energy and the frequency factor, calculated from the plot, are given in the lower part of Table 6.4. The relatively low value for E_a indicates that the rate-controlling step in the benzene production is probably the desorption of the product and the k's of Table 6.4 represent $k_s = k_4 + k'_{-4}$. This conclusion is consistent with that drawn earlier for the LaY catalyst: One of the two slow steps there must be the desorption of benzene. A second slow step is not observed in HY, and this is an indication that the dealkylation step is very fast here.

Table 6.4 Kinetic Parameters for Benzene Formation During Dealkylation of Cumene Over HY Catalyst

	$10^4 k$ (s^{-1})		Δk^a
T (K)	from F-peaks	from R-peaks	(%)
573	3.60 ± 0.07	3.8 ± 0.2	5.6
582	4.4 ± 0.7	4.8 ± 0.1	9.1
595	5.3 ± 0.1	5.6 ± 0.1	5.7
603	6.7 ± 0.2	6.4 ± 0.6	4.5
611	7.7 ± 0.1	7.2 ± 0.6	6.5
E_a (kJ mol^{-1}): 53 ± 3, ln (A/s^{-1}): 3.1 ± 0.5			

[a]Defined as $100|k_F - k_R|/k_F$.
Source: Ref. (5).

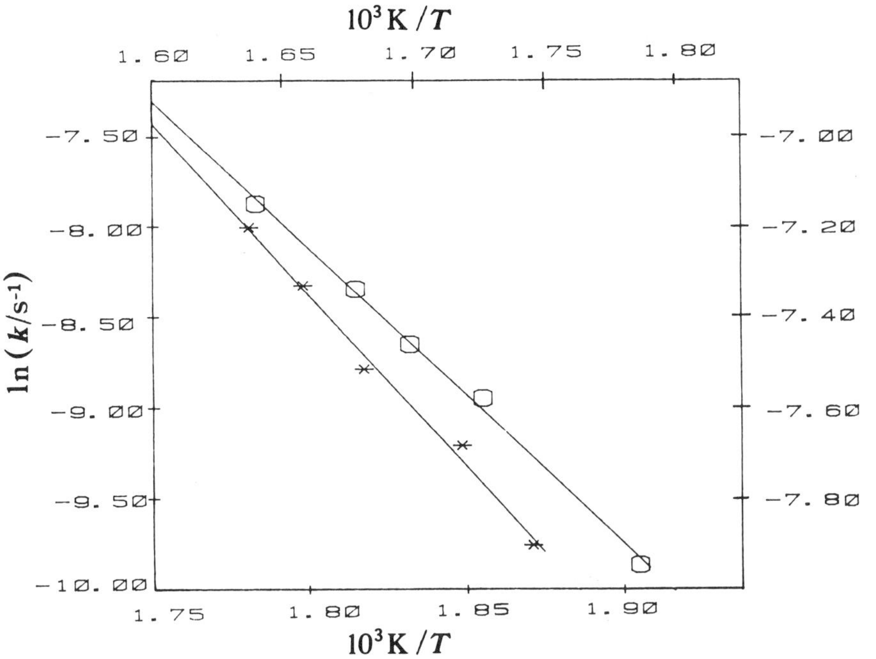

Fig. 6.13 Arrhenius plots for the cracking of cumene. O, Rate constants for formation of propene over LaY catalyst (lower abscissa, left ordinate); *, Rate constants for formation of benzene over HY catalyst (upper abscissa, right ordinate) (5).

It is seen from the foregoing discussion that the present method, applied to the catalytic cracking of cumene and based on the Campbell-Wojcienchowski mechanism, leads to an easy determination of rate constants, activation energies, and entropies of activation. The method uses simple integrated rate equations easily fitted to the experimental data, by means of which a limiting case of only one rate-controlling step can be distinguished from a case with two such steps.

IV. REACTION KINETICS WITH THE DIFFUSION TECHNIQUE

A. Theory

When a diffusion feed of the catalyst with the reactant(s) is used, as described in section II of this chapter, because these reactants are not retained for a long time by the catalytic bed the latter can be considered either as a differential reactor or treated in an integral mode. In the first case, the concentration of a product is given by the relation

$$c(l', t_0) = \frac{r(t_0)W}{\dot{V}} \tag{6-34}$$

where $r(t_0)$ is the mean reaction rate for the whole catalyst bed (mol s^{-1} g^{-1}), and W is the catalyst mass. The concentration of a reactant will depend not only on the reaction rate, but also on the rate of its diffusion into the carrier gas stream. It is calculated by solving the diffusion equation, Eq. (4-1):

$$\frac{\partial c_z}{\partial t_0} = D \frac{\partial^2 c_z}{\partial z^2} \tag{6-35}$$

under the appropriate initial condition and the necessary boundary conditions at both ends of the diffusion column L, as exemplified in chapter 4, section I.B. However, the boundary condition expressed by Eq. (4-4) must now take into account the chemical reaction at the exit of column L and be written as

$$D\left(\frac{\partial c_z}{\partial z}\right)_{z=0} = vc(l', t_0) + \frac{r(t_0)W}{a} \tag{6-36}$$

If the initial condition is that of Eq. (4-5), i.e., the reactant is introduced as a pulse at the closed end of column L, the solution of Eq. (6-35) with the boundary condition (6-36), carried out as before, is

$$c(l',t_0) = \frac{N_1 \exp(-L^2/4Dt_0)}{t_0^{3/2}} - \frac{r(t_0)W}{\dot{V}} \tag{6-37}$$

where N_1 is again given by Eq. (4-17).

If a relatively large volume of reactants is introduced into the diffusion column, so that a constant distribution in the length, L, can be assumed at $t_0 = 0$, the initial condition is $c_z(z) = c_0$, and the solution of the diffusion equation is [9]

$$c(l',t_0) = \frac{2D_2c_0}{vL_{eff}} \cdot \exp\left(-\frac{2D_2}{L^2_{eff}} \cdot t_0\right) - \frac{r(t_0)W}{\dot{V}} \tag{6-38}$$

where L_{eff} is equal to L when all diffusion length has the same temperature, or

$$L_{eff} = L_2 + L_1\left(\frac{D_2}{D_1}\right)^{\frac{1}{2}} \tag{6-39}$$

if only the lower part L_2 is inside the chromatographic oven at temperature T_2, whereas the upper part L_1 of the column is outside the oven at T_1. The D_2 and D_1 represent the diffusion coefficients of the reactant at T_2 and T_1, respectively.

If the diffusion feed is assumed to operate the catalytic bed in an integral way, it is preferable to use the total fractional conversion, x, of the reactant to product(s), instead of the concentration $c(l',t_0)$. This conversion is calculated experimentally from the sample peaks, either using their heights, h, or their area, f, under the curve, taking into account the relative response of the chromatographic detector for each substance. The equation relating the rate constant, k, with the fractional conversion of reactant to products is derived as follows (10). The specific reaction rate $r_m/\text{mol} \cdot \text{kg}^{-1}\ \text{cat} \cdot \text{s}^{-1}$ is

$$r_m = \frac{dx}{d\tau_m} = \frac{dx}{d(1/v_m)} = \frac{dx}{d(W/v)} \tag{6-40}$$

where τ_m is the space time of the reactor, v_m the reactor's space velocity (mol · kg^{-1} cat · s^{-1}), v the feed rate (mol s^{-1}), and W the catalyst's mass (kg).

Integration over the entire catalyst bed, taking into account the fact that during a flow reversal cycle the feed rate, v, remains practically constant, gives

$$\int_0^x \frac{dx}{r_m} = \int_0^W d\left(\frac{W}{v}\right) = \frac{W}{v} = \frac{W}{\dot{V}c_0} \qquad (6\text{-}41)$$

where $\dot{V}$ is the volume flow rate (m^3 s^{-1}) of the reacting mixture, and c_0 the extra concentration (mol m^{-3}) of the reactant in the mixture created by the flow reversal at x = l'; i.e., at the entry of the catalytic bed.

If there is only one reactant or the second reactant (like hydrogen in hydrogenation reactions) is used in great excess as carrier gas, only the order of the reaction, n, with respect to the first reactant is considered, so that the rate equation is of the form

$$r_m = kc^n = kc_0^{\,n}(1 - x)^n \qquad (6\text{-}42)$$

where k is the rate constant (m^{3n} mol^{1-n} kg^{-1} cat · s^{-1}). Substituting the right-hand side of Eq. (6-42) for r_m in Eq. (6-41) and integrating one obtains, after rearrangement

$$k = \frac{\dot{V}c_0^{\,1-n}}{W(1 - n)} \left[1 - (1 - x)^{1-n}\right] \qquad (6\text{-}43)$$

This equation holds for all values of n other than 1. In the latter case, the integration of Eq. (6-41) given (6)

$$k = \frac{\dot{V}}{W} \ln \frac{1}{(1 - x)} \qquad (6\text{-}44)$$

If the reaction is first order, the calculation of k by means of Eq. (6-44) requires only the fractional conversion, x, the weight of the catalyst, W, and the carrier gas flow rate, V, corrected at the catalytic bed temperature and pressure with the help of the relation

$$\dot{V} = \dot{V}_{meas} \cdot \frac{T_1}{T_2} \cdot \frac{p_2}{p_1} \tag{6-45}$$

where $\dot{V}_{meas}$ is the measured flow rate, and the subscripts 1 and 2 refer to the bed and the environment, respectively. However, for orders $n \neq 1$ Eq. (6-43) must be used, and this requires the additional quantity c_0; i.e., the concentration of the reactant at $x = l'$. This is not the total reactant concentration at this point, but only the extra concentration above the continuous baseline, created by the flow reversal and giving two sample peaks like those of Fig. 6.4; i.e., the measured conversion. According to the theory, the flow reversal for time, t', creates a square concentration function at $x = l'$ of width t' and height c_0 twice that corresponding to the existing baseline (cf. Fig. 3.6A). The area under this square sample peak is $c_0 t'\dot{V}$ mol and this should be proportional to the total area f_{tot} under the sample peaks as these are recorded by the detector system:

$$f_{tot} = f_r + Sf_p \tag{6-46}$$

where f_r and f_p are the peak areas for reactant and product, respectively, and S the response of the detector for the product relative to that of the reactant, under the experimental conditions used. Since, however, f_{t0t} is measured in mV s or in cm s, the sensitivity, S_c, in cm m^3 mol^{-1} of the detector is required to transform f_{tot} in mol, thus

$$c_0 t'\dot{V} = \frac{f_{tot}}{S_c} \dot{V} \text{ mol} \tag{6-47}$$

and if from this relation $f_{t0t}/S_c t'$ is substituted for c_0 in Eq. (6-43), we obtain

$$k = \frac{\dot{V}f_{tot}^{\,1-n}}{(1-n)W(S_c t')^{1-n}}\,[1-(1-x)^{1-n}] \tag{6-48}$$

The sensitivity, S_c, of the detector is easily found by injecting a known volume of reactant into column L and allowing it to diffuse into column l in the absence of catalyst. By making flow reversals and measuring the height, h(cm), of the sample peaks as a function of time, the diffusion coefficient of the reactant into the carrier gas is calculated as described in Chapter 4, section I. From the value of the intercept of the relevant plot the factor S_c is calculated.

In each kinetic run at a given temperature, the variables of Eq. (6-48) are f_{tot} and x. These are loaded into a computer together with the constants $\dot{V}$, W, and t' and the values of k are calculated by means of a suitable program for various values of n. Of these, we chose that n value which give the smallest variation in the k values. As a criterion for this variation are taken the 95% confidence limits of the mean k values.

B. Oxidation of Carbon Monoxide

In this reaction the cell of Fig. 6.14 was used. There are two gaseous reactants; i.e., carbon monoxide and oxygen, neither of which can be retained in the catalytic bed for a long time. It is necessary to use the diffusion technique, and a separate column of silica gel to separate the product, carbon dioxide, from the two reactants (7). The catalysts were Co_3O_4 supported on γ-Al_2O_3 doped with Ca^{2+} in varying amounts (25).

After conditioning the whole system and stabilizing the catalyst under carrier gas, helium, flowing from D_2 to D_1, 2 cm^3 of a 3 CO:1O_2 mixture at atmospheric pressure was rapidly introduced with a gas-tight syringe at the top of the diffusion column L.

The various concentrations at x = l' were sampled by reversing the flow for 12 s, which was smaller than both the gas hold-up time, t'_M, in the empty section l', and the retention time, t_R, of

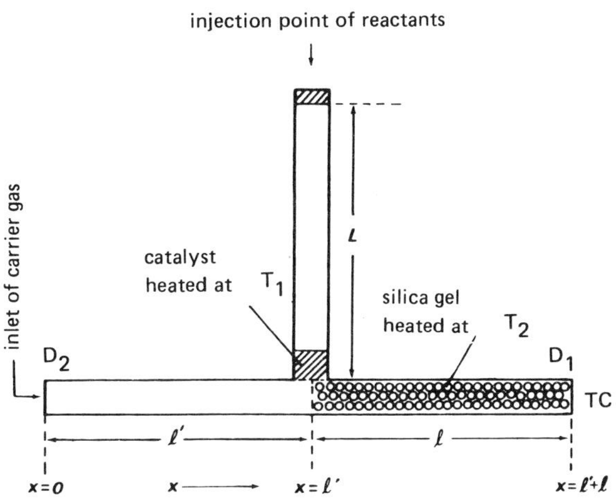

Fig. 6.14 Outline of the cell (with a thermal conductivity detector, TC) used to study the oxidation of carbon monoxide with oxygen gas over Co_3O_4 containing catalysts (Reproduced from Ref. 7 by permission).

all substances in column 1. Since t_R is different for the reactants $CO + O_2$ and the product CO_2, two fairly symmetrical peaks appear in each reversal, as shown in Fig. 6.15.

The catalytic bed was treated in a differential way, so that combining Eq. (3-27) for the product p:

$$h_p = 2c_p(l', t_0) \tag{6-49}$$

with Eq. (6-34), one obtains for the reaction rate

$$r(t_0) = \frac{h_p \dot{V}}{2W} \tag{6-50}$$

The concentration of the reactants, c_r, is also connected to the reactants' peak height, h_r, by an expression similar to Eq. (6-49). Their concentrations [CO] and $[O_2]$ can thus be calculated from h_r and h_p on the basis that: 1. steady-state conditions are established after a certain time, 2. the stoichiometry of the

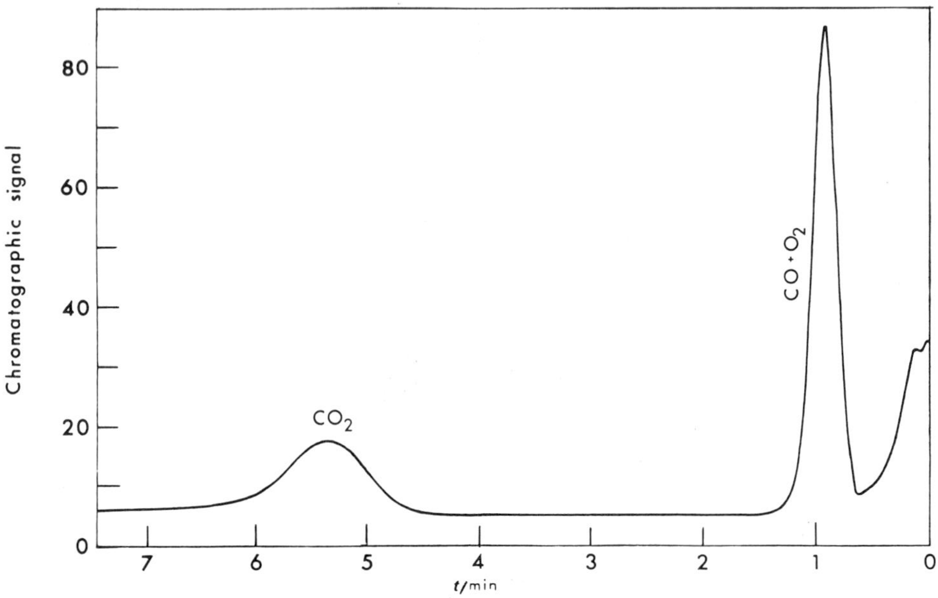

Fig. 6.15 Chromatographic sampling of the reaction $CO + \frac{1}{2}O_2 \rightarrow CO_2$ at 552.2°K, 44 min after injection of the reactants. The three arms of the cell of Fig. 6.14 had lengths L = 70.5 cm, l = l' = 40 cm and a common i.d. 4 mm. The catalytic bed contained 0.35 g catalyst (7) (Reproduced from Ref. 7 by permission).

reaction is $CO + 1/2\ O_2 \rightarrow CO_2$, 3. the initial composition of the reaction mixture was 3 $CO:1O_2$, and 4. The effective diffusion coefficients of CO and O_2 in He were found approximately equal. The above lead to the relations [7]

$$[CO] = \frac{3}{8} h_r + \frac{1}{16} h_p \qquad [O_2] = \frac{1}{8} h_r - \frac{1}{16} h_p \tag{6-51}$$

Finding [CO] and $[O_2]$ by Eq. (6-51) and $r(t_0)$ by Eq. (6-50), an apparent rate constant, k, can be computed for any chosen rate law, and then checked for constancy with time. Assuming linear Henry-type adsorption isotherms because of the small concentrations involved, a second-order rate law $r(t_0) = k[CO]\ [O_2]$

is expected for a simple Langmuir-Hinshelwood or Rideal mechanism. However, k values calculated using this rate law increase with time continuously without limit. This excludes a mechanism involving a reaction between adsorbed carbon monoxide and adsorbed or gaseous oxygen or vice versa. By contrast, another mechanism involving a reaction between adsorbed carbon monoxide and surface lattice oxygen (26) would follow a phenomenological rate law $r(t_0) = k[CO]$, if the mass of the catalyst is big enough. Rate constants, calculated on the basis of this rate law for all catalysts studied and at various temperatures, initially increase but then reach a limiting value and remain constant after a certain time (7). Activation energies can be calculated from this limiting k values in the usual way. They are given in Table 6.5.

C. Hydrogenation of Propene

The experimental arrangement of Fig. 6.3, with silica gel as chromatographic material filling column 1' + 1 and no additional separation column, was used for the catalytic hydrogenation of propene

Table 6.5 Activation Energies for the Oxidation of Carbon Monoxide over Co_3O_4 Supported on γ=Al_2O_3 Doped with Ca^{2+}, for Various Support Compositions

Ca^{2+} content (mmol/g γ-Al_2O_3)	E_a (kJ mol^{-1})
0.392	33 ± 5
0.621	24 ± 7
0.984	14 ± 1
1.560	26 ± 1
2.470	44 = 6

± values are standard errors from the regression analysis
Source: Ref (7), with permission.

over 5% platinum on alumina (9). The reaction was studied at temperatures considerably higher (409-439°K) than those (275-308°K) in the classic study of Rogers et al. [27]. At such temperatures, the conversion of propene to propane is so high that a detailed kinetic study would be very difficult by the conventional methods. Equal volumes of the reactants C_3H_6 and H_2 were premixed, and 10 cm^3 of this mixture was slowly introduced in column L, in each kinetic run. Therefore, Eqs. (6-34) and (6-38) for propane and propene, respectively, must be used to express $c(l',t_0)$, in conjunction with Eq. (3-27) giving the height of the sample peaks. An example of sampling this reaction is shown in Fig. 6.16. Clearly, from the relative peak heights of propane and propene (corrected for possibly different sensitivity of the detector), their relative concentrations at the reaction site can be revealed at any chosen time, rendering easy the test of various kinetic equations and reaction orders. As a first example, consider the equation

$$r(t_0) = \frac{kp_H p_A}{1 + K_A p_A}$$

or

$$\frac{p_H p_A}{r(t_0)} = \frac{1}{k} + \frac{K_A}{k} \cdot p_A \tag{6-52}$$

where p_H and p_A are partial pressures of H_2 and alkene, respectively. This equation was found to fit the kinetic data of ethene hydrogenation over a supported nickel catalyst (28). Let us examine if Eq. (6-52) describes the results of propene hydrogenation obtained here. Using Eqs. (3-27), (6-34), and (6-38), and the stoichiometry of the reaction $A + H_2 \rightarrow B$, where A stands for propene and B for propane, we express $r(t_0)$, p_A, and p_H as follows:

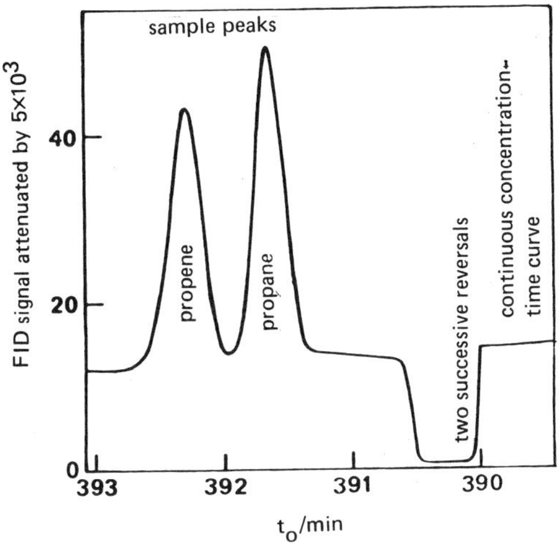

Fig. 6.16 A chromatographic sampling of the gases involved in the propene hydrogenation over platinum supported on alumina. It shows the sample peaks of the product propane and the unreacted propene. The temperature was 438.9°K, and the inert carrier gas nitrogen at a volume flow rate $\dot{V} = 0.350$ cm^3 s^{-1}. A flame ionization detector was used (9).

$$r(t_0) = \dot{V}h_B/2W \tag{6-53}$$

$$p_A = RT[c_A(l',t_0) + c_B(l',t_0)] = RT\,\frac{h_A + h_B}{2} \tag{6-54}$$

$$p_H = RT[c_H(l',t_0) + c_B(l',t_0)] = RT\,\frac{2D_H c_0}{vL_{eff}} \cdot \exp\left(-\frac{2D_H}{L^2_{eff}}\cdot t_0\right) \tag{6-55}$$

Substituting these relations in Eq. (6-52) and rearranging one obtains

$$\frac{h_A + h_B}{h_B} \cdot \exp\left(-\frac{2D_H}{L^2_{eff}} \cdot t_0\right) = \frac{1}{RT\omega k} + \frac{K_A}{2\omega k}(h_A + h_B) \tag{6-56}$$

where

$$\omega = 2RT \frac{WD_H c_0}{v\dot{V}L_{eff}} \tag{6-57}$$

The left-hand side of Eq. (6-56), containing only experimentally determined quantities, was found to decrease with time continuously, whereas $h_A + h_B$ on the right-hand side remained practically constant for more than 5 h. These findings clearly show that Eq. (6-52) is not consistent with the experimental data. Another noticeable fact is that $(h_A + h_B)/h_B$, which is simply the reciprocal of the fractional conversion of propene to propane, increases with time and then reaches a limiting constant value. Therefore, the empirical dependence of the reaction rate on the concentrations of the reactants must have the form

$$r(t_0) = k_1[C_3H_6][H_2]^n + k_2[C_3H_6] \tag{6-58}$$

where k_1 and k_2 are apparent rate constants. Substituting Eq. (6-53) for $r(t_0)$, $(h_A + h_B)/2$ for $[C_3H_6]$, and $(2D_H c_0/vL_{eff})$ $\exp(-2D_H t_0/L^2_{eff})$ for $[H_2]$ as Eqs. (6-54) and (6-55) imply, we obtain after some rearrangement

$$\frac{h_B}{h_A + h_B} = \frac{k_1 W}{\dot{V}}\left(\frac{2D_H c_0}{vL_{eff}}\right)^n \cdot \exp\left(-\frac{2nD_H}{L^2_{eff}} \cdot t_0\right) + \frac{k_2 W}{\dot{V}} \tag{6-59}$$

The left-hand side of this equation is the fractional conversion, x, of propene to propane, and the last term, $k_2W/\dot{V}$, on the right-hand side is the conversion at infinite time, x_∞. Eq. (6-59) can therefore be written in linear form as

$$\ln (x - x_{\infty}) = \ln Q - (2nD_H/L^2_{eff})t_0 \tag{6-60}$$

where Q signifies the preexponential factor of Eq. (6-59). An example of plotting the experimental data according to this equation is given in Fig. 6.17. From the slope $-2nD_H/L^2_{eff}$ the order, n, of the reaction with respect to hydrogen can be found, using the known values of D_H and L_{eff} (cf. Eq. (6-39)). Alternatively, the value of $2D_H/L^2_{eff}$ at the temperature of each run can be determined experimentally in the absence of catalyst or in the absence of propene, by plotting the logarithm of the sample peak height of hydrogen versus t_0, according to Eq. (6-38). The experimental value of $2D_H/L^2_{eff}$ in the absence of catalyst and with column l' + l filled with silica gel, combined with Eq. (6-39) and $D_2/D_1 = (T_2/T_1)^{3/2}$, gave a value for D_H which differed from the theoretically calculated value by only 1.7%. In the presence of 50 mg of catalyst, however, a value for D_H 31.1% bigger than before was found. This is explained by slow adsorption of hydrogen

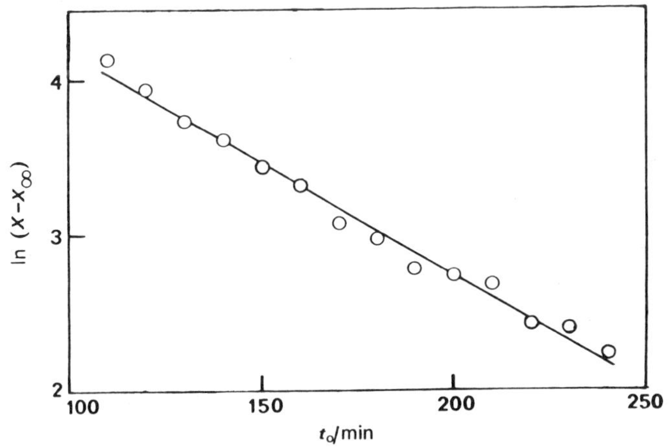

Fig. 6.17 Conversion data for hydrogenation of propene over platinum supported on alumina, at 428.5°K, plotted according to Eq. (6-60). The carrier gas was nitrogen (9).

on the catalyst surface, and it is probably the reason for the dependence of the reaction order n(1.6-2.5) on the hydrogen pretreatment. The importance of this pretreatment was also noted by Rogers et al. (27).

D. Hydrodesulfurization of Thiophene

The reversed-flow technique using a diffusion feed has also been applied to the industrially important reaction of catalytic hydrodesulfurization of thiophene over cobalt-molybdenum oxides supported on four types of commercial alumina (6). The experimental setup is that depicted in Fig. 6.7, with the reactant thiophene (0.5 cm^3 of liquid) placed in the upper section of the diffusion column (80 cm × 2 mm i.d.). The separation column (6 m × 2 mm i.d.) was filled with 20% Carbowax, 20 M, on Chromosorb P, and kept at a constant temperature of 395°K. Hydrogen was used as carrier gas and as second reactant.

The simple flow reversals (from R to F direction) give rise to various positive and negative frontal boundaries and some sample peaks, as illustrated in Fig. 6.18. These are explained as follows. With the hydrogen flowing in direction R, the separation column and the branch l of the sampling cell (cf. Fig. 6.7) contain reactant and products with concentrations $c^{(r)}$ and $c^{(p)}$, respectively, whereas the branch l' is empty of any substance. After time $t_R^{(p)}$ (which is the retention time of the products in the separation column) from the reversal of the carrier gas flow to the F direction, a negative frontal boundary appears with height $h^{(p)} \simeq c^{(p)}$. For time t_M', which is the gas hold-up time in the branch l', the elution curve is due only to the reactant still remaining in the separation column, but after this time new products appear, contained in the sampling branch l and passing through the catalyst. These products create the peak P_1. After time $t_R^{(r)}$ (which is the retention time of the reactant thiophene in the separation column) a new negative frontal boundary appears with

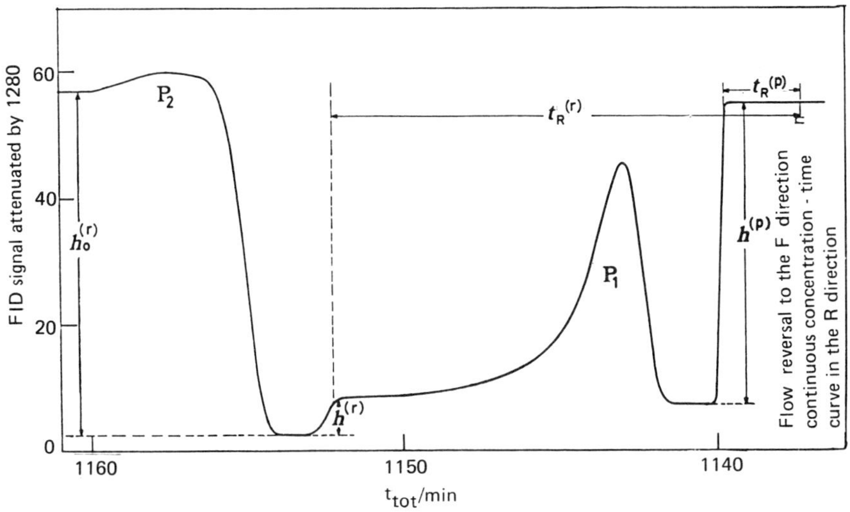

Fig. 6.18 Frontal boundaries and sample peaks obtained by reversing the carrier gas flow from R- to F-direction, during the hydrodesulfurization of thiophene over Co_3O_4—MoO_3—Al_2O_3 catalyst, at 507°K (6).

height $h^{(r)}$, which measures the concentration of thiophene after passing through the catalyst bed. The elution curve then remains at the zero point for time t'_M, after which a new sample peak, P_2, appears superimposed on the ending baseline. This baseline has a height $h_0^{(r)}$, which measures the concentration of thiophene before its passing through the catalyst bed.

A single reversal of the carrier gas flow from R to F direction gives the total fractional conversion of the reactant to products:

$$x = \frac{h_0^{(r)} - h^{(r)}}{h_0^{(r)}} \qquad (6\text{-}61)$$

This sampling process is repeated many times in the catalyst's steady state, giving various values of x with a small standard deviation from each other (6). This constancy of x with time indicates that inhibition of the reaction by H_2S produced from thiophene is negligible. The absence of inhibition by H_2S is not unexpected because the catalysts were sulfided before use. In view of the above, first-order kinetics with respect to thiophene can be assumed, as has also been done by others (29).

Treating x calculated by Eq. (6-61) as integral conversion, one can find the rate constant k using Eq. (6-44). To calculate actual specific reaction rates, r_m, at the entry of the bed, it is only necessary to multiply the rate constant by the concentration, c_0, of thiophene. The latter can be found from the characteristics of the reactant feeding arrangement (cf. Fig. 6.7), as has been done in another study (30).

The Arrhenius plots for the catalysts used are given in Fig. 6.19. With the exception of the catalyst supported on alumina Houdry Ho 417, the remaining three catalysts give two straight lines on these plots. This is not a rare phenomenon (31) in the catalytic reactions and can be due to two reasons. First, to diffusion phenomena, and a chemical reaction being diffusion limited at relatively high temperatures. Second, to the small surface coverage at high temperatures, lowering the true activation energy by the heat of adsorption of thiophene ΔH_a (a negative quantity) and giving an apparent activation energy:

$$E_a^{app.} = E_a^{true} + \Delta H_a \tag{6-62}$$

Activation energies and frequency factors for two temperature ranges (lower and higher than 500°K) are given in Table 6.6. The E_a listed show that at temperatures lower than about 500°K, the rate-determining step is the chemical reaction on the catalyst surface, whereas at temperatures higher than 500°K, the E_a values appear smaller than those expected (except for the catalyst supported on alumina Houndry Ho 417). If this were due to diffusion limitations, the apparent E_a values would be expected in the region

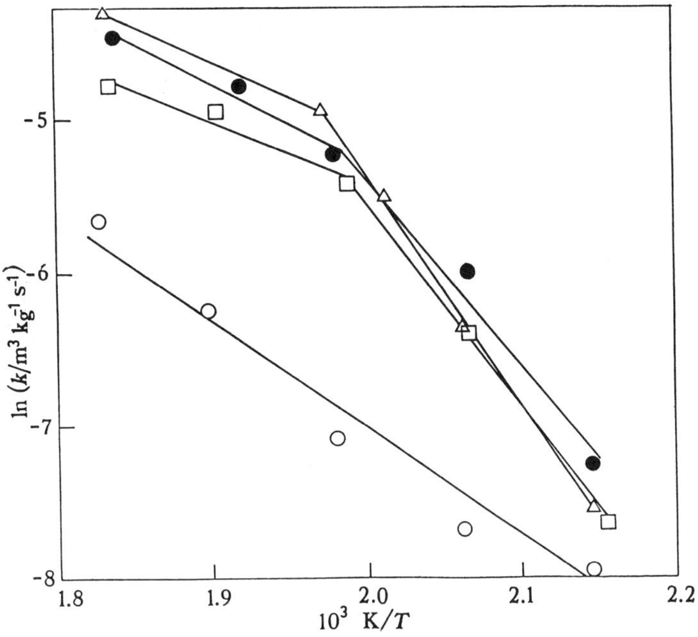

Fig. 6.19 Arrhenius plots for the hydrodesulfurization of thiophene over Co_3O_4 — MoO_3 — Al_2O_3 catalysts. (O) Alumina Ho 417; (□) alumina Ho 415; (△) alumina Ho 425; (•) alumina commercial Greek. The two latter plots have the ordinate shifted down by 0.4 units; i.e., the -5 should be read -5.4, etc. (6).

of 10 — 15 kJ mol^{-1}. Since the values found at high temperatures are all higher than 35 kJ mol^{-1}, the heat of adsorption of thiophene must be involved at high temperatures.

Thus from Eq. (6-62), the heats of adsorption can be calculated as:

$$\Delta H_a = E_a^{app} - E_a^{true} = E_a^{high\ T} - E_a^{low\ T} \qquad (6\text{-}63)$$

These values are also given in Table 6.6.

A comparison of the catalytic activity as determined by this method with that found by other researchers using different

Table 6.6 Activation Energies (E_a) and Frequency Factors (A) for Two Temperature Ranges Together with Heats of Adsorption (ΔH_a) for the Hydrodesulfurization of Thiophene Over Four Catalysts.

	E_a (kJ mol^{-1})		ln A		
Alumina	T <500°K	T >500°K	T <500°K	T >500°K	ΔH_a (kJ mol^{-1})
Houdry Ho 415	110	35	21.0	3.0	75
Houdry Ho 417	62	-	7.9	-	-
Houdry Ho 425	125	38	24.3	3.6	87
Commercial Greek	101	44	18.6	4.9	57

Source: Ref. (6).

techniques is not an easy task, since most workers simply measure fractional conversions of the reactant to products and no kinetic parameters as was done here. Owens and Amberg [32] have found an apparent activation energy for the hydrodesulfurization of thiophene over cobalt-molybdate catalyst of 104.6 kJ mol^{-1}. This value is in fairly good agreement with the activation energies determined here (cf. Table 6.6) for the three most active catalysts at temperatures lower than 500°K.

Thomas et al. [29], studying the hydrodesulfurization of thiophene over a catalyst of 18.6% MoO_3 on γ-Al_2O_3, have found a rate constant at 673°K 1.05×10^{-3} m^3 thiophene converted · (mol metal)$^{-1}$ s^{-1}. Multiplying this value by 1.2922 mol metal/kg cat. gives -1.3568×10^{-3} m^3 thiophene converted · kg^{-1} cat · s^{-1}. Dividing the last number by 0.062, since the gas phase contained 6.2 mol% thiophene in hydrogen, the rate constant k comes out as 2.2×10^{-2} m^3 · kg^{-1} cat · s^{-1} at 673°K. In the present instance, the rate constant for the catalyst prepared from the Greek alumina at 673°K (calculated by using the Arrhenius parameters of Table 6.6) is 5.1×10^{-2} m^3 · kg^{-1} cat · s^{-1}; i.e., only 2.3 times larger than that of Thomas et al. (29).

The final conclusion of this study is that the detailed kinetics of catalytic hydrodesulfurization reaction, which is very important for the refined oil industry, can be studied with simplicity and accuracy using the new RFGC technique.

E. Methanation of Carbon Monoxide

Finally, the diffusion feed technique has been used to study the methanation reaction of carbon monoxide over Ni/Al_2O_3 catalysts, employing the experimental arrangement of Fig. 6.6, with hydrogen as carrier gas and a thermal conductivity detector (10). All three branches L, l', and l, of the reversed-flow cell were of stainless-steel chromatographic tube with an inside diameter of 4 mm. The diffusion column was empty and had a length of 100 cm. Sections l' and l were each 115 cm long and were both filled with chromatographic material, except a short length (5mm) of l near the junction with L which contained 53 or 107 mg of 5 or 2% (w/w) Ni/Al_2O_3, 80-100 mesh catalyst, respectively. This part of column l was heated with a separate heating element and its temperature was measured with a separate thermocouple. The variations during each run were less than 1°K.

The pressure drop along column l' was the same as that along column l. It was measured with a mercury manometer and found that the catalyst bed was under a pressure of 2.015 atm during all experiments.

Conditioning of the catalyst was carried out in situ at 713°K for 16 h, under carrier gas (H_2, 0.75 cm^3 s^{-1}) flowing from D_2 to D_1 (cf. Fig. 6.6) and with 30 cm^3 CO at atmospheric pressure injected with a gas-tight syringe into column L. The temperature of the chromatographic oven was 346 ± 1°K in all experiments.

Following the above conditioning, and after the chromatographic signal having been decayed to a negligible height, a new 30-cm^3 volume of CO at atmospheric pressure was slowly injected into the column L. After about 10 min, a continuous concentration-time curve decreasing slowly is established in the recorder owing to both the reactant CO and the products. This can be taken as the steady-state condition for the catalyst. Then, the carrier gas flow is reversed in direction by turning the six-port valve from one position (solid lines) to the other (dashed lines). After 30 s of backward flow, the carrier gas is turned again to the original direction. Because the time t' elapsing between two successive reversals of the flow is less than the retention time on columns l' and l of the substances giving rise to the chromatographic signal, a symmetrical sample peak for each substance follows the restoration of the gas flow to its original direction, as shown in Fig. 6.4. The above procedure of the flow reversals was repeated many times at each temperature, giving rise to a series of sample peaks which correspond to different partial pressures of CO over the catalyst and different fractional conversions of reactants to products.

The pressure change in tube L and over the catalyst at each flow reversal was negligibly small owing to the same pressure drop along columns l' and l and to the time interval, t', being short. We must remind that the height of each sample peak above the continuous chromatographic signal (taken as baseline) is proportional to the concentration of the substance giving rise to this peak, at the junction x = l' (cf. Fig. 6.3) and at the time of the flow reversal. In the present instance, however, this would be true in the absence of catalyst, while with the catalytic bed present at the entrance of column section l, the sample peaks at the detector, like those of Fig. 6.4, are due to substances present at the exit of the bed; i.e., after passage of the carbon monoxide sample peak at x = l' through the catalyst. The only product detectable in measurable amounts was methane. The water expected as a second product was presumably retained on the molecular sieve 5A used as chromatographic material in columns l' and l. By using Chromosorb 102 and silica gel in place of molecular sieve 5A, the formation of carbon dioxide, ethane, and higher paraffins was sought, but none of them was observed in detectable amounts.

The catalytic fractional conversion, x, of reactants to products was calculated from the methane and the carbon monoxide sample peaks detected after each flow reversal. The areas f under the sample peaks were used instead of their heights because owing to the different retention times of CO and CH_4 on column l the two peaks had different widths at their half heights. Therefore, the conversion, x, was calculated by the relation

$$x = \frac{0.76 f_{CH_4}}{0.76 f_{CH_4} + f_{CO}} \qquad (6\text{-}64)$$

0.76 being the response of the thermal conductivity detector for

CH_4 relative to that for CO, under the experimental conditions used. By applying Eq. (6-46) one finds

$$f_{tot} = F_{CO} + 0.76\ f_{CH_4} \tag{6-65}$$

and having determined the sensitivity of the detector, S_c, for CO, as explained in section IV. A, we can use Eq. (6-48) to calculate the order of the reaction n and the rate constant k.

Activation parameters are determined from Arrhenius plots, using in each plot rate constants pertaining to the same reaction order. Four such plots are given in Fig. 6.20, corresponding to orders -0.5, -0.2, +0.2, and +0.5. These plots show that activation energies and frequency factors remain constant within the range of temperatures of the same reaction order. Table 6.7 collects the activation parameters found.

The order of the reaction n with respect to carbon monoxide has been reported may times (33-35) to change with the partial pressure of CO, and with the temperature. In the present work, the order n changed with temperature, but not with the concentration c_0 of CO in the gas mixture at the same temperature. In ech kinetic run c_0 changed about tenfold, but this led to a single value for the rate constant k and a single value for the order n. The simplest explanation of this lies probably in what has been pointed out under item 5 of the list in section II.B.2. As regards the change of n with temperature, the most noticeable finding is its <u>discontinuous</u> change, whereas in the literature (33,35) a continuous change is reported in about the same temperature ranges. Moreover, the usual findings have been that n is negative, increases with temperature, and tends to zero at high ttemperature, whereas here positive values of n are found at high temperatures. The results are consistent with the following rather conventional reaction mechansim, assuming a nonhomogeneous catalytic surface with independent adsorption sites • and * for H and CO, C, O, CH_x species, respectively:

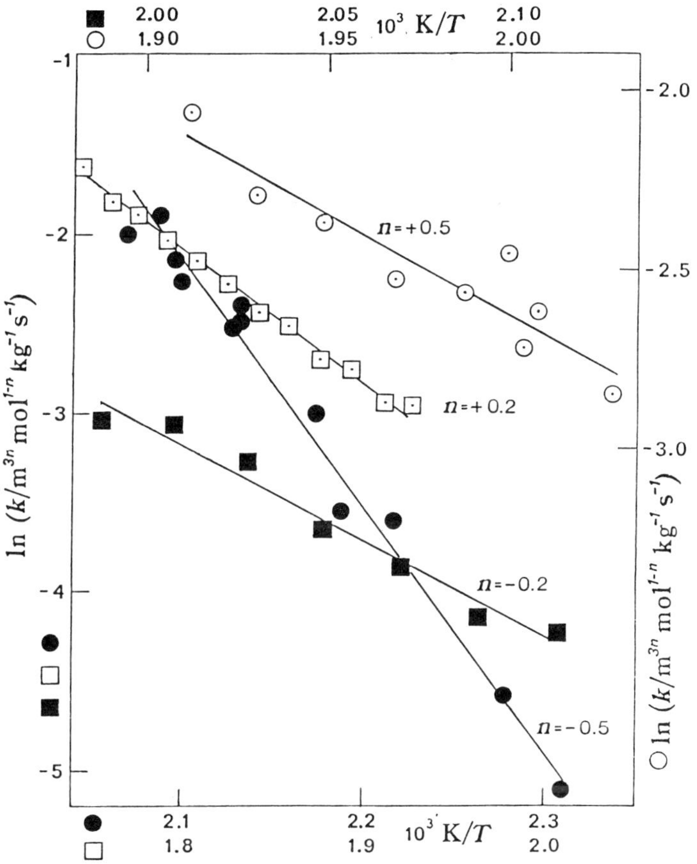

Fig. 6.20 Arrhenius plots for the methanation of carbon monoxide over Ni/Al_2O_3 catalysts. The four plots correspond to different reaction orders, n, with respect to CO. The n = -0.5 plot includes two -0.45 and one -0.55 points; the n = +0.5 plot includes two points with order +0.60 and one with +0.40; the n = -0.2 plot includes one point with order -0.24 and one with -0.16. The various symbols O, •, ▫ and ▪ refer to the coordinate and the abscissa scales on which they are written (10).

Table 6.7 Activation Energies E_a and Frequency Factors A for the Methanation of Carbon Monoxide over two Ni/Al_2O_3 Catalysts

Catalyst	n	E_a (kJ mol^{-1})	ln (A/m^{3n} mol^{1-n} kg^{-1} cat. s^{-1})
5% (w/w)Ni/Al_2O_3	-0.5	112 ± 5	26 ± 1
	+0.5	47 ± 7	9 ± 2
2% (w/w)Ni/Al_2O_3	-0.2	89 ± 8	18 ± 2
	+0.2	62 ± 2	11.2 ± 0.4

± values are standard errors.
Source: Ref. (10).

$$\mathrm{CO(g)} + * \overset{K_1}{\rightleftharpoons} \mathrm{CO}*$$

$$\mathrm{CO}* + * \overset{K_2}{\rightleftharpoons} \mathrm{C}* + \mathrm{O}*$$

$$\mathrm{H_2(g)} + 2\bullet \overset{K_3}{\rightleftharpoons} 2\mathrm{H}\bullet$$

$$\mathrm{C}* + \mathrm{H}\bullet \overset{K_4}{\rightleftharpoons} \mathrm{HC}* + \bullet$$

$$\mathrm{HC}* + \mathrm{H}\bullet \xrightarrow{k_1} \mathrm{H_2C}* + \bullet$$

$$\mathrm{H_2C}* + \mathrm{H}\bullet \overset{K_5}{\rightleftharpoons} \mathrm{H_3C}* + \bullet$$

$$\mathrm{H_3C}* + \mathrm{H}\bullet \xrightarrow{k_2} \mathrm{CH_4(g)} + * + \bullet$$

$$\mathrm{O}* + 2\mathrm{H}\bullet \xrightarrow{k_3} \mathrm{H_2O(g)} + * + 2\bullet \qquad (6\text{-}66)$$

where $K_1 - K_5$ are equilibrium and $k_1 - k_3$ rate constants.

Steady-state kinetics based on the above mechanism lead to the following equation for the rate of formation of CH_4 (10):

$$r_{CH_4} = \frac{k' c_{H_2}}{\left(1 + K_3^{1/2} c_{H_2}^{1/2}\right)^{3/2}} \cdot \frac{c_{CO}^{1/2}}{1 + K' c_{CO}^{1/2} + K_1 c_{CO}} \qquad (6\text{-}67)$$

where c_{H_2} and c_{CO} are gas phase concentrations (partial pressures p_{H_2} and p_{CO} could be used as well), and k', K' are given by the relations

$$k' = (k_1 k_3 K_1 K_2 K_4)^{1/2} K_3 \qquad (6\text{-}68)$$

$$K' = \left(\frac{k_3 K_1 K_2}{k_1 K_4}\right)^{1/2} \cdot \left(\frac{1}{1 + K_3^{1/2} c_{H_2}^{1/2}}\right)^{1/2} \qquad (6\text{-}69)$$

Since H_2 is used as carrier gas in all experiments and the gas concentration of CO in it is relatively low, the first fraction on the right-hand side of Eq. (6-67) can be considered as constant k" at constant temperature, and by further division of both terms of the remaining fraction by $c_{CO}^{1/2}$ Eq. (6-67) becomes

$$r_{CH_4} = \frac{k''}{c_{CO}^{-1/2} + K' + K_1 c_{CO}^{1/2}} \tag{6-70}$$

The reaction order as a function of temperature, starting from a negative value of n at low temperatures, passing through zero, and ending with the respective positive value at high temperatures can now easily be explained with the help of Eq. (6-70). According to this rate equation, the apparent experimental order of the reaction depends on the relative magnitudes of the three terms in the denominator. At low temperatures, the equilibrium constant K_1 for the CO adsorption may become big enough to make the term $K_1 c_{CO}^{1/2}$ predominate over the other two terms, $c_{CO}^{-1/2}$ and composite constant K', which can be neglected. This limiting case leads then to an order $n = -1/2$ with respect to c_{CO}. As Eq. (6-69) shows, K' consists of two factors; i.e., a ratio of rate and equilbirium constants which may decrease as the temperature rises, and a hydrogen adsorption factor $1/(1 + K_3^{1/2} c_{H_2}^{1/2})^{1/2}$ which increases with increasing temperature. It is possible that as the temperature rises, K' increases, passes through a maximum, and then decreases. At the temperature of the maximum K' value, this would predominate over the other two terms in the denominator of Eq. (6-70), giving zero-order kinetics. Finally, at high enough temperatures, both K' and $K_1 c_{CO}^{1/2}$ may become negligible compared to $c_{CO}^{-1/2}$, thus leading to an order $n = +1/2$.

Since the apparent rate constant is composed of different constants in each reaction order, as Eq. (6-70) shows, activation energies and frequency factors should remain constant within the temperature range where the reaction order is kept constant, and should change when the order changes. This is in accord with the results of Table 6.7.

As a conclusion, we can say that the new technique of RFGC used here reveals characteristics of the reaction not observed with other methods. This is due most probably to what has been pointed out earlier, i.e., it is simultaneously a pulse and a continuous feed method, extracting all the information expected from a pulse technique under steady-state conditions.

V. MASS TRANSFER COEFFICIENTS OF VAPORS ON SOLIDS

Mass transfer phenomena can be studied, not only with a filled diffusion column L and an empty sampling column l' + l, as was discussed in Chapter 5, but also the other way round; i.e., an empty column L and a filled column l' + l. Virtually all mass transfer studies in GC were made at infinite solute dilution by using conventional elution of an injected pulse. No mass transfer coefficient measurements at finite solute concentration (nonlinear region of chromatography) have come to our attention. These would obviously require the use of some kind of frontal analysis, displacement development, or special equipment for finite concentration work. However, the new technique of reversed-flow gas chromatography requires only a slight modification of a conventional gas chromatograph, and can lend itself to mass transfer measurements at finite concentration, using elution bands like those obtained with trace amounts of solute.

For this application (8), a porous polymer (Porapak P) was chosen, because such materials are widely used today as sorbent phases in gas-solid chromatography, and also because relatively few quantitative data on solute mass transfer in gas-solid systems are available.

An experimental setup similar to that used to measure gas diffusion coefficients (Fig. 4.1) was employed with a slight modification, as shown in Fig. 6.21.

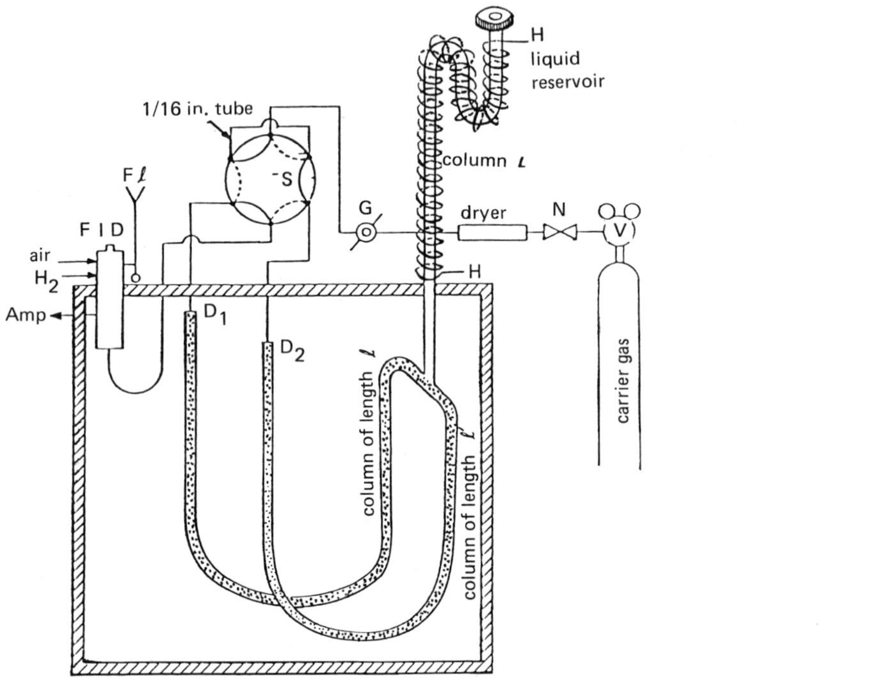

Fig. 6.21 Gas lines and connections for measuring mass transfer coefficients by RFGC. V = two-stage reducing valve and pressure regulator; N = needle valve; H = heating coil by water circulation; G = gas flow controller; S = six-port valve with a short 1/16-in. tube connecting two alternate ports; Fl = bubble flowmeter; Amp = signal to amplifier (8).

The sections l' + l of the sampling column (39 + 39 cm × 4 mm i.d. were filled with Porapak P. A diffusion column L (80 cm × 4 mm i.d.) empty of any chromatographic material and having a U-shaped liquid reservoir at its upper end was connected perpendicularly at its lower end to the middle of the filled column l + l'. The column L and the liquid reservoir were kept at a constant temperature in the range 48-62°C by means of circulating water around them from a thermostat.

The conditioning of the chromatographic column l + l' containing the Porapak P was carried out at 170°C for 24 h with a carrier gas flowrate 0.33 $cm^3\ s^{-1}$. After that the column is brought to each working temperature, and while carrier gas is flowing in a certain direction through the column l + l', a small amount of liquid solute (usually 0.5 cm^3) is introduced by injection into the heated upper deposit of column L. After a certain time, during which no signal is noted, an ascending concentration-time curve for the solute is recorded. This reaches a maximum plateau and remains there as long as there is still liquid in the reservoir (see Fig. 5.7). Thus a finite solute concentration in the Porapak column is established, the magnitude of which depends on the vapor pressure of the liquid solute at the temperature of the reservoir. When sufficient stability of the recorded signal is attained, valve S is switched to the other position (dashed lines), thereby reversing the direction of the carrier gas flow. After a short time interval (20-60 s) of backward flow, t', the valve is again turned to its previous position, thus restoring gas flow to the original direction.

Because the duration t' of backward flow is smaller than the retention time of the solute t_R and t_R' on column sections l and l', respectively, it creates a perturbation in the concentration-time line having the form of a bell-shaped sample peak. This emerges with a characteristic retention time following the restoration of the gas flow to the original direction, and "sits" on the otherwise finite concentration signal (cf. Fig. 6.22). The procedure is repeated several times, giving a series of sample peaks.

The pressure drop along column l or l' is found by measuring the pressure at the injection point of the solute with an open mercury manometer.

The mass transfer resistance coefficients, C, together with the coefficients $\bar{B}$, are found by fitting the experimental data to the simple classic van Deemter equation $\hat{H} = A + \bar{B}/\bar{v} + C\bar{v}$,

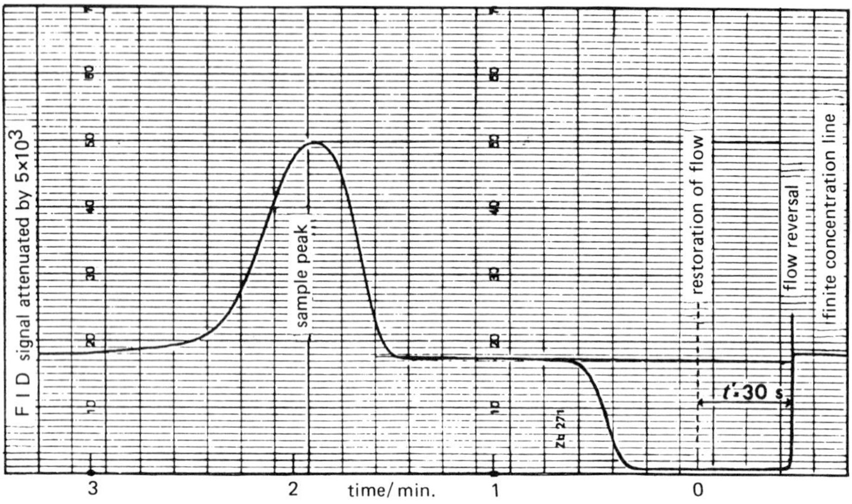

Fig. 6.22 A sample peak created by reversing the flow direction of carrier gas for 30 s while a finite concentration of n-heptane was passing through the column at 408°K, and a volumetric flow-rate of 1.07 $cm^3\ s^{-1}$ (8).

using a least-squares program on a suitable computer. The apparent (i.e., experimental) plate height, $\hat{H}$, is calculated from the relation $\hat{H} = l/N$, and the number of theoretical plates, N, from the ratio t_R^2/σ^2_{net}, where t_R is the retention time of the sample peak, measured from the moment of the flow restoration to the original direction, and σ^2_{net} the net variance of the peak. The two latter experimental parameters are computed as follows.

According to the theory of the RFGC method, the reversal of the flow for a time, $t' < t_R$, creates a plug (i.e., a square function) on the continuous signal due to the diffusion of vapors from column L into column $l + l'$. This square function has a theoretical width t', i.e., equal to the time of the backward flow, and its variance is therefore $t'^2/12$. When passed through the Porapak column section l or l', the sample peak will be given by the difference of two error functions (36). Its variance, σ_{exp}^2, and retention time, t_R, is calculated by the rectangular method of peak area measurement introduced by Sternberg (37). Finally, the net variance due to the chromatographic process is found from the difference

$$\sigma_{net}^2 = \sigma_{exp}^2 - \frac{t'^2}{12} \tag{6-71}$$

Table 6.8 summarizes the results obtained as outlined above. Before discussing the mass transfer term C, it is worthwhile to comment on the values of A and $\bar{B}$ coefficients. The particle diameter for a 80-100 mesh solid material, like that used here, is 0.018-0.015 cm and most A values in Table 6.8 are of that order of magnitude, as in most cases at infinite dilution studies. The negative values are probably due to experimental errors, since the term A is very sensitive to these errors.

By using the Fuller-Schettler-Giddings equation (38), the diffusion coefficients, D_m, of the solute vapors in the carrier gas helium were calculated at each column temperature. These (given also in Table 6.8 for 1 atm pressure) were used to find the obstruction factor γ from the well-known relation $\bar{B} = 2\gamma D_m$. In this calculation, each D_m value is first reduced to the mean column pressure, corresponding to the mean flow velocity of the carrier gas. The γ values found, listed in the last column of Table 6.8, seem reasonable.

Coming to the C coefficients, we first note that they do not include a mobile phase diffusion term C_m, since they are independent of pressure. There are three main points to be discussed: 1. the magnitude of C terms, 2. their variation with t', i.e., with the width of the square function created by the flow reversal, and 3. their variation with temperature. As regards point 1, the C coefficients found are too large to be attributed to adsorption-desorption kinetics, i.e., being $C_k = 2k/(1 + k)^2 k_d$,

Table 6.8 Values for the Coefficients A, $\bar{B}$ and C of Van Deemter Equation Determined by RFGC at Finite Solute Concentration

Solute	T (K)	t' (s)	A (cm)	$\bar{B}$ ($cm^2\ s^{-1}$)	10^2 C (s)	D_m ($cm^2\ s^{-1}$)	γ
n-Heptane	393	30	0.029	0.271	1.34	0.443	0.39
	393	60	-0.004	0.289	3.40	0.443	0.39
	408	20	0.058	0.253	1.40	0.462	0.32
	408	30	0.079	0.225	1.85	0.462	0.29
	408	40	-0.015	0.376	2.98	0.462	0.51
	423	20			2.76		
	423	30	0.041	0.294	4.95	0.503	0.33
	423	40	-0.179	0.544	8.90	0.503	0.64
	438	30	0.002	0.264	5.40	0.535	0.29
n-Hexane	423	30	-0.139	0.410	9.80	0.515	0.49
n-Octane	423	30			1.15		

Source: Ref. (8).

where k is the partition ration and k_d the desorption rate constant. The C_k terms are of the order of 10^{-8}-10^{-6} s, and therefore their contribution to the magnitude of C found must be negligible. We can therefore adopt the explanation that it is surface diffusion which controls the mass transfer of solute in the solid phase.

Coming now to point 2 above, i.e., the variation of C with the reversal time t', Table 6.8 shows that there is a 2.1- to 3.2-fold increase of C by doubling t'. This can be explained by referring to the finite solute concentration, on account of which the isotherm may not be linear. A decrease in the isotherm slope brought about by an increase in t' will cause a decrease in k. Since C is usually of the form

$$C = N \frac{k}{(1 + k)^2 D_{eff}} \tag{6-72}$$

where N is a constant and D_{eff} an effective diffusion coefficient pertaining to the appropriate mass transfer mechanism, C will increase with decreasing k (provided that $k > 1$) and hence with increasing t'.

Finally, the variation of C with temperature for t' constant (30 s) is shown in Fig. 6.23 in the form of a plot of ln C versus 1/T. It shows an increase of C with temperature, and this is an unusual dependence, at least for infinite dilution studies. An obvious explanation for this lies on the decrease of k with temperature. For $k \gg 1$, Eq. (6-72) becomes

$$C \simeq \frac{N}{kD_{eff}} = \frac{N'}{KD_{eff}} = N'' \exp\left[\frac{\Delta H_{ad} + E}{RT}\right] \tag{6-73}$$

where N' and N" are constants, K is the adsorption equilibrium constant, ΔH_{ad} the heat of adsorption, and E the activation energy of D_{eff}. According to this equation, a plot of ln C versus 1/T should be linear with a slope equal to $(\Delta H_{ad} + E)/R$. From

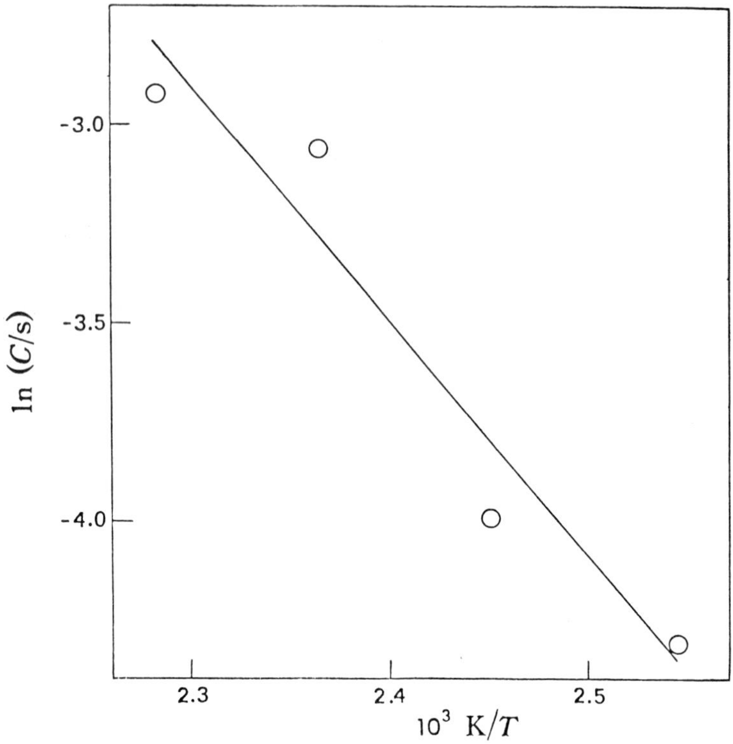

Fig. 6.23 Temperature dependence of C for n-heptane on a Porapak P column (8).

Fig. 6.23 we find a slope of -5.931.10^3, and this gives ΔH_{ad} + E = -49.3 kJ mol^{-1}. The heat of vaporization for n-heptane at its normal boiling point (98.4°C) is 31.7 kJ mol^{-1}, and if we take E = 0, because of the small activation energy expected, the heat of adsorption is only 1.6 times bigger than the heat of vaporization. This places the adsorption of n-heptane on Porapak P in the domain of physical adsorption. For comparison we quote here the heat of adsorption of n-heptane on graphitized carbon black, equal to -52.3 kJ mol^{-1}.

LIST OF SYMBOLS

a	Volume of gas phase per unit length of column, or cross-sectional area of void space
A, B	Kinetic parameters defined by Eqs. (6-24) and (6-25), respectively
c, c_z	Concentration of a solute vapor in the sampling column and the diffusion column, respectively
c_0	Initial concentration of a reactant, extra concentration of a reactant created by flow reversal
C	Laplace transform of c with respect to t_0, mass transfer resistance coefficient
C_z	Laplace transform of c_z with respect to t_0
D	Mutual diffusion coefficient of two gases
f	Area under the curve of sample peaks
g	Fraction of reactant on reactive sites of surface
h	Height of sample peaks measured from the ending baseline
k	Rate constant, partition ratio
k_1, k_1', k_2, k_3, k_4, k_{-1}, k_{-2}, k_{-3}, k_{-3}'	Rate constants
k_a	Apparent rate constant defined by Eq. (6-30)
k_s	Sum of k_3 and k_{-3}'
K_1, K_2, K_3, K_4	Equilibrium constants
K_A, K_D	Adsorption equilibrium constants (partition coefficients) of substances A and D
l, l'	Lengths of the two sections of the sampling column
L, L_1, L_2	Diffusion lengths
L_{eff}	Effective diffusion length defined by Eq. (6-39)
m	Amount of a reactant injected
n	Order of reaction
N_1	Parameter defined by Eq. (4-17)
p_0	Transform parameter with respect to t_0

Q, Q_{abs}	Parameter defined by Eqs. (6-11) and (6-14), respectively
r_m	Specific reaction rate
$r(t_0)$	Reaction rate at t_0
$R(p_0)$	Laplace transform of $r(t_0)$ with respect to t_0
S_c	Sensitivity of the detector
t_0	Time from the beginning to the last backward reversal of gas flow
t'	Time interval of backward flow
t_{tot}	Total time equal to $t_0 + t'$
t_R, t'_R	Ideal retention time on the filled column section l or l', respectively
τ	Time defined by Eq. (3-20)
τ_m	Reactor's space time
v	Linear velocity of carrier gas, feed rate
v_m	Reactor's space velocity
V	Volume of gas phase in the column
$\dot{V}$	Volume flowrate of carrier gas
x	Distance coordinate in column $l' + l$, fractional conversion of reactants to products
W	Catalyst's mass
ω	Parameter defined by Eq. (6-57)

REFERENCES

1. N. A. Katsanos and I. Georgiadou, J. Chem. Soc., Chem. Commun., 242 (1980).
2. N. A. Katsanos, J. Chem. Soc., Faraday Trans I, 78:1051 (1982).
3. G. Karaiskakis, N. A. Katsanos, I. Georgiadou, and A. Lycourghiotis, J. Chem. Soc., Faraday Trans I, 78:2017 (1982).
4. M. Kotinopoulos, G. Karaiskakis, and N. A. Katsanos, J. Chem. Soc., Faraday Trans. I, 78:3379 (1982).
5. N. A. Katsanos and M. Kotinopoulos, J. Chem. Soc., Faraday Trans. I, 81:951 (1985).
6. N. A. Katsanos, G. Karaiskakis, and A. Niotis, J. Catal., 94:376 (1985).

7. G. Karaiskakis, N. A. Katsanos, and A. Lycourghiotis, Can. J. Chem., 61:1853 (1983).

8. E. Dalas, G. Karaiskakis, N. A. Katsanos, and A. Gounaris, J. Chromatogr., 348:339 (1985).

9. N. A. Katsanos, G. Karaiskakis, and A. Niotis, Proc. 8th Int. Congr. Catal., Vol. III, Dechema-Verlag Chemie, West Berlin, 1984, p. 143.

10. E. Dalas, N. A. Katsanos, and G. Karaiskakis, J. Chem. Soc., Faraday Trans. I, 82:2897 (1986).

11. G. Karaiskakis and N. A. Katsanos, Proc. 3rd Mediterranean Congr. Chem. Eng., Barcelona, 1984, p. 68.

12. S. J. Gentry and R. Rudham, J. Chem. Soc., Faraday Trans. I, 70:1685 (1974).

13. D. A. Best and B. W. Wojciechowski, J. Catal., 47:343 (1977).

14. C. D. Prater and R. M. Lago, in Advances in Catalysis (D. D. Eley, W. G. Frankenburg, V. I. Komarewski, and P. B. Weisz, eds.), Vol. 8, Academic Press, New York, 1956, p. 293.

15. D. R. Campbell and B. W. Wojciechowski, J. Catal., 20:217 (1971).

16. D. A. Best and B. W. Wojciechowski, J. Catal., 31:74 (1973).

17. D. Best and B. W. Wojciechowski, J. Catal., 47:11 (1977).

18. D. A. Best and B. W. Wojciechowski, Prepr.-Can. Symp. Catal., 5:400 (1977).

19. D. A. Best and B. W. Wojciechowski, J. Catal., 53:243 (1978).

20. D. A. Best and B. W. Wojciechowski, Can. J. Chem. Eng., 56:588 (1978).

21. A. Corma and B. W. Wojciechowski, Prepr.-Can. Symp. Catal., 6:149 (1979).

22. A. Corma and B. W. Wojciechowski, J. Catal., 60:77 (1979).

23. A. Corma and B. W. Wojciechowski, Can. J. Chem. Eng., 58:620 (1980).

24. A. Corma, H. Farag., and B. W. Wojciechowski, Int. J. Chem. Kinet., 13:883 (1981).

25. A. Lycourghiotis, A. Tsiatsios, and N. A. Katsanos, Z. physik. Chem., 126:95 (1981).

26. D. Mehanddjiev and E. Nikolova-Zhecheva, J. Catal., 65:475 (1980).

27. G. B. Rogers, M. M. Lih, and O. A. Hougen, A. I. Ch. E. J., 12:369 (1966).

28. Ho-Feng Koh and R. Hughes, J. Catal., 33:7 (1974).

29. R. Thomas, E. M. van Oers, V. H. J. de Beer, J. Medema, and J. A. Moulijn, J. Catal., 76:241 (1982).

30. G. Karaiskakis and N. A. Katsanos, J. Phys. Chem., 88:3674 (1984).

31. R. M. Massagutor et. al., Paper presented at 7th World Petroleum Congress, Mexico City, 1967.

32. P. J. Owens and C. H. Amberg, Can. J. Chem., 40:491 (1962).

33. J. A. Dalmon and G. A. Martin, J. Catal., 84:45 (1983).

34. J. Klose and M. Baerns, J. Catal., 85:105 (1984).

35. M. A. Vannice, Catal. Rev. Sci. Eng., 14:153 (1976).

36. J. C. Sternberg, Adv. Chromatogr., 2:260 (1966).

37. J. C. Sternberg, Adv. Chromatogr., 2:215 (1966).

38. E. N. Fuller, P. D. Schettler, and J. C. Giddings, Ind. Eng. Chem., 58:19 (1966).

Index

Activity coefficients, 198-202
Adsorption chromatography, 2,3
Adsorption-desorption kinetics, 2, 4, 32, 71, 72, 74, 289
Adsorption isotherms, 2,4
Affinity chromatography, 3
Air
  critical volume of, 149
  diffusion coefficient of, 146
  molecular diameter of, 148
  relative molar response for, 147
Alcohols, dehydration of, 239-243
Amines, deamination of, 243-247
Argon
  critical volume of, 149
  diffusion coefficient of, 146
  molecular diameter of, 148
  relative molar response for, 147
Arrhenius equation, 46-47, 48, 193
Binary mixtures, diffusion coefficients in, 113-138
Broadening factors in gas chromatography, 17-27
  longitudinal diffusion 18-23
  noninstantaneous equilibration of solute between two phases, 23-27
n-Butane, enthalpy and entropy of adsorption of, 211
Butan-l-ol, mass transfer coefficient of, 190
1-Butene
  critical volume of, 149
  diffusion coefficient of, 146
  molecular diameter of, 148
  relative molar response for, 147
cis-2-Butene
  critical volume of, 149
  diffusion coefficient of, 146
  molecular diameter of, 148
  relative molar response for, 147
trans-2-Butene
  critical volume of, 149
  diffusion coefficient of, 146

[trans-2-Butene]
molecular diameter of, 148
relative molar response for, 147
But-1-ene, enthalpy and entropy of adsorption of, 211

Carbon dioxide
critical volume of, 149
diffusion coefficient of, 146
molecular diameter of, 148
relative molar response for, 147
Carbon monoxide
methanation of, 277-285
oxidation of, 264-267
Carrier gases, 3
diffusion coefficients of five solutes in, 132-133
diffusion coefficients of three solutes in, 134-135
flowrate perturbation, 34-35
Catalytic conversions calculated by stopped-flow gas chromatography, 67-71
Catalytic cracking of cumene, 247-259
Chromatographic sampling equation, 95-105
Chromatography, historical background of, 1-2
Collision diameters, 142-153
Compensation effect, 31-32
Cracking, catalytic, of cumene, 247-259
Critical volumes of gases, 142-153
Cumene, catalytic cracking of, 247-259
Deamination of primary amines, 243-257
Dehydration of alcohols, 239-243
Diffusion coefficients, 18, 27, 35, 77, 79, 80, 88, 113, 115, 116, 121, 122, 128
in binary mixtures, 113-138
of five solutes in three carrier gases, 132-133
of gases in carrier gas helium, 146
measurement of, 77-81
in multicomponent gas mixtures, 138-142
of three solutes in two carrier gases, 134-135
of vapors, 180-202
in helium, 190
Diffusion column
empty, reversed-flow with, 113-161
diffusion coefficents in multicomponent gas mixtures, 138-142
gas diffusion coefficients in binary mixtures, 113-138
Lennard-Jones parameters, 150, 151, 154-158
list of symbols, 158-160
relative molar responses, collision diameters, and critical volumes of gases, 142-153
filled, reversed-flow with, 163-224
list of symbols, 221-222
mass transfer across a gas-liquid boundary, 165-180
mass transfer across phase boundaries extended to all the diffusion column, 212-221

[Diffusion column]
mass transfer across phase boundaries near the junction of the diffusion and the sampling column, 203-212
mass transfer coefficients for the evaporation of liquids and diffusion coefficients of vapors, 180-202
Diffusion technique, 227, 229, 230-236
advantages of, 234-236
reaction kinetics with, 260-285
hydrodesulfurization of thiopene, 272-277
hydrogenation of propene, 267-272
methanation of carbon monoxide, 277-285
oxidation of carbon monoxide, 264-267
theory, 260-264
Displacement development, 4, 5-7
Double Laplace transforms, 20, 56, 61, 98, 119

Einstein's law of diffusion, 22
Elution chromatogram, 16
Elution development, 4-5
Enthalpies of adsorption of hydrocarbons, 211
Entropies of adsorption of hydrocarbons, 211
Equations of gas chromatography, 7-29
broadening factors, 17-27
experimental determination of retention volumes, 14-16
[Equations of gas chromatography]
retention parameters, 7-14
theoretical plates and van Deemter equation, 27-29
Equilibrium chromatography (ideal chromatography), 4
Equilibrium constants, calculation from sample peaks, 106-108
Ethane
critical volume of, 149
diffusion coefficient of, 146
enthalpy and entropy of adsorption of, 211
molecular diameter of, 148
relative molar response for, 147
Ethanol, mass transfer coefficient of, 190
Ethene
critical volume of, 149
diffusion coefficient of, 146
enthalpy and entropy of adsorption of, 211
molecular diameter of, 148
relative molar response for, 147

First-order conservation equation, 7
Flick's second law, 18
Flowrate perturbation, 34-35
Frontal analysis, 4,5

Gas absorption, mass transfer across a gas-liquid boundary and, 165-180
Gas chromatography, 1-40

[Gas chromatography]
basic equations of, 7-29
broadening factors, 17-27
experimental determination of retention volumes. 14-16
retention parameters, 7-14
theoretical plates and van Deemter equation, 27-29
carrier gas flowrate perturbation, 34-35
chromatographic methods, 4-7
list of symbols, 36-38
physicochemical measurements by, 30-34
terminology, 2-4
Gas-liquid chromatography (GLC), 2, 3
Gas-solid chromatography (GSC), 2, 3
Gel-permeation chromatography, 3

Henry's law constant, 4, 170, 176, 177, 179
n-Heptane, mass transfer coefficient of, 190
n-Hexane, mass transfer coefficient of, 190
Hydrocarbons, enthalpies and entropies of adsorption of, 211
Hydrodesulfurization of thiopene, 272-277
Hydrogenation of propene, 267-272

Ideal chromatography (equilibrium chromatography), 4
Ideal stopped-flow gas chromatography, 48-71
catalytic conversions calculated by stopped-flow GC, 67-71
general theoretical analysis, 49-57
nonopposing reaction on one kind of active site with no intermediate, 58-61
nonopposing reactions on two kinds of active sites, 61-65
nonopposing reactions on three kinds of active sites, 65-67
Inverse Laplace transforms, 12, 20, 24, 53, 57, 60, 101, 124, 182
Ion-exchange chromatography, 3

Kinetics
adsorption-desorption, 2, 4, 32, 71, 72, 74, 289
reaction, with diffusion technique, 260-285
hydrodesulfurization of thiopene, 272-277
hydrogenation of propene, 267-272
methanation of carbon monoxide, 277-285
oxidation of carbon monoxide, 264-267
theory, 260-264
reaction, with pulse technique, 236-259
catalytic cracking of cumene, 247-259
deamination of primary amines, 243-257
dehydration of alcohols, 239-243

[Kinetics]
theory, 236-239
of sulfur dioxide action on marble, 216-221
of surface-catalyzed reactions, 226-236

Langmuir's adsorption isotherm, 4
Laplace transforms, 11, 12, 19, 20, 21, 23, 52, 53
double, 20, 56, 61, 98, 119
inverse, 12, 20, 24, 53, 57, 60, 101, 124, 182
Lennard-Jones parameters, 150, 151, 154-158
Linear chromatography, 4
Linear-ideal chromatography, 4
Liquid chromatography, 2,3
Liquid-liquid chromatography, (LLC), 3
Liquid-solid chromatography, 3

Marble, kinetics of sulfur dioxide action on, 216-221
Mass transfer
across a gas-liquid boundary, 165-180
across phase boundaries extended to all the diffusion column, 216-221
across phase boundaries near the junction of the diffusion and the sampling column, 203-212
Mass transfer coefficients, 4, 27, 88, 168, 169, 176, 177, 179
for the evaporation of liquids, 180-202
pure liquids, 190
[Mass transfer coefficients]
for liquid mixtures, 193-198
of vapors and solids, 285-292
Methanation of carbon monoxide, 277-285
Methane, enthalpy and entropy of adsorption of, 211
Methanol, mass transfer coefficient of, 190
2-Methylpropene
critical volume of, 149
diifusion coefficient of, 146
molecular diameter of, 148
relative molar response for, 147
Mobile phase, 3
Molecular diameters of various gases, 148

Nitrogen
critical volume of, 149
diffusion coefficient of, 146
molecular diameter of, 148
relative molar response for, 147
Non-ideal chromatography, 4
Non-ideal stopped-flow gas chromatography, 71-77
Nonlinear chromatography, 4
Nonlinear gas chromatography, 30-31
Nonpposing reactions
on one kind of active site, 58-61
on two kinds of active sites, 61-65
on three kinds of active sites, 65-67

n-Octane, mass transfer coefficient of, 190

Oxidation of carbon monoxide, 264-267
Oxygen
  critical volume of, 149
  diffusion coefficient of, 146
  molecular diameter of, 148
  relative molar response for, 147

Paper chromatography, 2,3
Partition chromatography, 2, 3
Partition coefficient, 10, 14, 16, 23, 71, 169, 176, 177, 179, 205
Partition ratio, 10
n-Pentane, mass transfer coefficient of, 190
Pentan-1-ol, mass transfer coefficient of, 190
Physicochemical measrements, 30-34
Propane
  critical volume of, 149
  diffusion coefficient of, 146
  molecular diameter of, 148
  relative molar response for, 147
Propan-1-ol, mass transfer coefficient for, 190
Propene
  critical volume of, 149
  diffusion coefficient of, 146
  enthalpy and entropy of adsorption of, 211
  hydrogenation of, 267-272
  molecular diameter of, 148
  relative molar response for, 147
Pulse technique, 227, 236
  reaction kinetics with, 236-259
    catalytic cracking of cumene, 247-259
[Pulse technique]
    deamination of primary amines, 243-247
    dehydration of alcohols, 239-243
    theory, 236-239

Rate coefficients, calculation of, from sample peaks, 106-108
Reaction kinetics
  with the diffusion technique, 260-285
    hydrodesulfurization of thiopene, 272-277
    hydrogenation of propene, 267-272
    methanation of carbon monoxide, 277-285
    oxidation of carbon monoxide, 264-267
    theory, 260-264
  with the pulse technique, 236-259
    catalytic cracking of cumene, 247-259
    deamination of primary amines, 243-247
    dehydration of alcohols, 239-243
    theory, 236-239
Relative molar responses (RMR), 142-153
Retention parameters, 7-14
Retention volumes, 14-16
Reversed-flow technique, 35, 87-111
  calculation of rate coefficients and equilibrium constants from sample peaks, 106-108
  chromatographic sampling equation, 95-100
  general experimental setup,

[Reversed-flow technique]
88-92
list of symbols, 108-109
reversed-flow GC as a sampling technique, 92-95
Reversed-flow with empty diffusion column, 113-161
diffusion coefficients in multicomponent gas mixtures, 138-142
gas diffusion coefficients in binary mixtures, 113-138
Lennard-Jones parameters, 150, 151, 154-158
list of symbols, 158-160
relative molar responses, collision diameters, and critical volumes of gases, 142-153
Reversed-flow with filled diffusion column, 163-224
list of symbols, 221-222
mass transfer across a gas-liquid boundary, 165-180
mass transfer across phase boundaries extended to all the diffusion column, 212-221
kinetics of action of sulfur dioxide on marbles, 216-221
mass transfer across phase boundaries near the junction of the diffusion and the sampling column, 203-212
mass transfer coefficients for the evaporation of liquids and the diffusion coefficients of
[Reversed-flow with filled diffusion column]
vapors, 180-202
Reversed-flow with filled sampling column, 225-296
kinetics of surface-catlyzed reaction, 226-236
list of symbols, 293-294
mass transfer coefficients of vapors on solids, 285-292
reaction kinetics with the diffusion technique, 260-285
hydrodesulfurization of thiopene, 272-277
hydrogenation of propene, 267-272
methanation of carbon monoxide, 277-285
oxidation of carbon monoxide, 264-267
theory, 260-264
reaction kinetics with the pulse technique, 236-259
catalytic cracking of cumene, 247-259
deamination of primary amines, 243-247
dehydration of alcohols, 239-243
theory, 236-239

Sample peaks, calculation of rate coefficients and equilibrium constants from, 106-108
Sampling column, filled, reversed-flow with, 225-296
kinetics of surface-catalyzed reactions, 226-236
list of symbols, 293-294
mass transfer coefficients of

[Sampling column, filled, reversed-flow with]
vapors on solids, 285-292
reaction kinetics with the diffusion technique, 260-285
hydrodesulfurization of thiopene, 272-277
hydrogenation of propene, 267-272
methanation of carbon monoxide, 277-285
oxidation of carbon monoxide, 264-267
theory, 260-264
reaction kinetics with the pulse technique, 236-259
catalytic cracking of cumene, 247-259
deamination of primary amines, 243-247
dehydration of alcohols, 239-243
theory, 236-239
Stationary phase, 3
Stefan-Maxwell equations, 141-142
Stopped-flow technique, 35, 41-85
experimental setup, 47-48
ideal stopped-flow gas chromatography, 48-71
catalytic conversions calculated by stopped-flow
[Stopped-flow technique]
GC, 67-71
general theoretical analysis, 49-57
nonopposing reaction on one kind of active site with no intermediate, 58-61
nonopposing reactions on two kinds of active sites, 61-65
nonopposing reactions on three kinds of active sites, 65-67
limitations of, 87-88
list of symbols, 81-83
measurement of diffusion coefficients in gases, 77-81
nonideal stopped-flow gas chromatography, 71-77
Sulfur dioxide, kinetics of action on marble of, 216-221
Supercritical fluid chromatography, 3

Theoretical plates, 27-29
Thin-layer chromatography, 2, 3
Thiopene, hydrodesulfurization of, 272-277

Van Deemter equation, 27-29